T0258874

Introduction to
Static Analysis Using
SolidWorks Simulation®

Introduction to
Static Analysis Using
SolidWorks Simulation®

Radostina V. Petrova

CRC Press
Taylor & Francis Group
Boca Raton   London   New York

CRC Press is an imprint of the
Taylor & Francis Group, an **Informa** business

CRC Press
Taylor & Francis Group
6000 Broken Sound Parkway NW, Suite 300
Boca Raton, FL 33487-2742

© 2015 by Taylor & Francis Group, LLC
CRC Press is an imprint of Taylor & Francis Group, an Informa business

No claim to original U.S. Government works

Printed on acid-free paper
Version Date: 20140514

International Standard Book Number-13: 978-1-4822-3618-7 (Hardback)

Visit the Taylor & Francis Web site at
http://www.taylorandfrancis.com

and the CRC Press Web site at
http://www.crcpress.com

To my loving family

CONTENTS

FOREWORD

This book on static analysis using the SolidWorks Simulation® tool is written to give a practical problem-based introduction in the use of a finite element simulation approach within a computer-aided design (CAD) tool environment. Nowadays, finite element analysis (FEA) is becoming a versatile approach to analyse complex structures. Contrary to earlier approaches where computer-aided tools were on their own isolated islands of automation, performing design, analysis, simulation and other computerised techniques within a single environment has been found beneficial for several reasons. As a result, we find strong collaborations among developers of today's computer-aided engineering (CAE) tools.

SolidWorks is one of the advanced and widely used CAD tools in use both in academia and in the industry. Convinced by the benefits of incorporating simulation at early design stage where a designer tests, optimises and simulates the real-world situation without developing costly prototypes, SolidWorks Simulation provides a user-friendly virtual design and prototyping environment. Though the general concept of design simulation using numerical methods is advanced, this book presents an approach where a user can simulate his/her design and gets the feeling of the functionality without deep knowledge of the numerical calculations behind the simulation tool. At the same time, the book attempts to give the basics of the working principles and analysis steps of numerical simulation approaches in general within the simulation examples executed in the book. Therefore, it is the author's belief that, upon reading the book, the user or the reader gets not only an idea how to use SolidWorks design and simulation functions but also a sufficient level of understanding the working principles of the numerical calculations and the conditions under which the user can make a successful simulation.

The special features of the book are that the user is guided by step-by-step procedures and graphical tools are extensively used to aid easier access to the functions in the software. In addition, key action words are written in bold text. These are mainly intended particularly for new users so that getting used to the graphical user interface and the functionality of the tools is simplified, and the learning curve of new users becomes steep.

The design and simulation principles discussed in the book are further demonstrated in a separate but accompanying solutions manual. Based on the selected 14 case studies, this book attempts to illustrate design and simulation principles for both simpler and relatively complex cases.

Hirpa G. Lemu, PhD
Associate Professor of Mechanical Design Engineering
University of Stavanger
Norway

PREFACE

This book is intended to help students and graduates in their first attempts to develop a static analysis of a structure using SolidWorks Simulation®. Complementary, the book can benefit professionals who have initial training in finite element method and are accustomed to the basics of solid mechanics.

The book adopts the SolidWorks software for conducting finite element analysis (FEA) because it is one of the most widely used software packages in mechanical design and related fields. Its features are explained through solving a set of industrial examples, showing different case studies and discussing the impact of the selected options on the result.

After reading the book, students and professionals can independently test their newly acquired knowledge by solving the examples in the attached solution manual.

The development of CAD models is not the focus of the book, but it is a prerequisite for successful understanding of the given samples. Therefore, the readers can either establish the 3D models of the examples themselves, following the instructions in the book and in the solution manual.

The language of the book is easy to follow, granted there are many technical terms; but given the subject, this is inevitable. Any terms that may not be familiar to a practicing engineer or to an engineering student are explained in a way appropriate for undergraduates with little software skills and for inexperienced software users. The adopted 'step-by-step' approach, combined with extensive explanatory notes, figures and icons, benefits the understanding of what is being done and guides the readers straight to the next level of performing the FEA. The taught material and what the reader should have learned are summarised after each chapter. Thus, the readers can easily track and assess their progress.

Finally, providing all this knowledge, the book outlines the path that the readers can follow to implement correct and reasonable static analysis, and sets the foundation of their professional improvement in the CAD/CAE field.

Models and images created in this text utilise SolidWorks® and SolidWorks Simulation®. SolidWorks is a registered trademark of Dassault Systemes SolidWorks Corporation, Waltham, MA, USA.

ACKNOWLEDGMENTS

This book might not have been possible without the strong support of Dr. Gagandeep Singh, senior commissioning editor for Engineering and Environment Sciences at CRC Press; Mrs. Stephanie Morkert, project coordinator at Taylor & Francis, LLC, who guided my first steps as an author and helped me throughout the entire process of writing this book; Mrs. Marie Planchard, director of education community, SolidWorks, who encouraged me; and my colleagues and friends, who convinced me to share my knowledge and experience.

Last but not least, I would like to thank my family for their patience and love.

AUTHOR

Radostina Petrova has a MSc Eng degree in civil engineering – structural design (calculations) of industrial and residential buildings. She has been working as a structural engineer for few years. Since 2007, she has been a self-employed licensed professional building engineer.

Dr. Petrova received her PhD degree in applied mechanics from the Technical University of Sofia, Bulgaria. In 2003, she was awarded an Ernst Mach research grant for young scientists by the Ministry of Youth, Science, and Education of Austria and adopted the grant at the Vienna University of Technology, Austria, investigating the oscillation of a bi-cable aerial ropeway under lateral wind excitation. She was awarded research grants under the Financial Mechanism of European Economic Area and conducted investigations on the dynamics of a horizontal wind turbine (in 2012) and a robot for medical (surgery) operation (in 2014) at the University of Stavanger, Norway.

In 2007, Dr. Petrova was appointed as associate professor in dynamics, strength and reliability of machines, devices and systems at the Technical University of Sofia.

Dr. Petrova has been recognized as an expert by the Research Executive Agency, Brussels, Belgium; by the National Centre for Research and Development of Poland; and by the Ministry of Education and Science of Bulgaria.

Her research interests and fields of expertise include multi-body dynamic simulation of mechanical systems; nonlinear structural analysis; structural modelling and analysis using FEM; simulation-based, design optimisation of mechanical systems; CAD/CAE (FEA) design of structures and mechanical systems, particularly dynamic analysis and simulations; structural engineering; wind engineering; fluid–structure interaction; exposure of slender structures (aerial ropeways, wind turbines, etc.) to random dynamic excitation; and interaction and combination of different software platforms/data for solving different structural problems.

CHAPTER 1

INTRODUCTION

1.1 OBJECTIVES OF THE BOOK

The objective of this book is to introduce the basic features of SolidWorks (SW) Simulation through solving a few practical examples. Therefore, we will start our course with trying to answer two main questions:

- Why use the finite element method (FEM)? What are its advantages and disadvantages compared to other numerical methods?
- Why have we chosen SW? Can we use any other software to obtain similar results?

At this very moment, you have to trust and agree with my reasons, but I believe that by the time you finish reading this book, you will have enough knowledge and experience to make your own choice and to find the answers to the above questions. Even more, after reading the book, you should have a good understanding of the logic of finite element analysis (FEA) and the obligatory stages you have to perform when using whatever software to adopt the FEM.

1.2 BASICS CONCEPTS OF FEM

There are many numerical methods for modelling, analysing, and simulating different engineering systems or processes. The earliest sources of publications related to FEM could be traced back to the mid-1960s; however, the FEM became popular some decades later with the invention and improvement of computers and the necessary software, and its rise continues up to the present. As used by the modern engineers, FEM represents the confluence of three ingredients: Matrix Structural Analysis (MSA), variation approximation theory, and the digital computer.

Nowadays, the FEM is one of the most widely used techniques for standard designs of engineering objects due to its generality and suitability for computer implementation. Due to the existence of a large amount of software based on FEM concepts and their easy adoption by users with different levels of experience, this software can be found in almost every design bureau, industry department, vocational school and technical university. There is no need to be an expert in the details of the FEM to solve common engineering problems and to handle everyday design tasks.

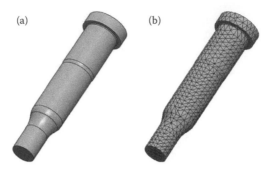

Figure 1.1
Models of a hole punch. (a) CAD model of a part. (b) FE model of the same part.

The principle of the FEM is to slice the solid body into many small, simply shaped cells, which would model the geometry of the body as accurately as possible (Figure 1.1). These small cells are called **finite elements** (FEs) or simply elements. They contact each other at points called **nodes**. The process of transformation of the solid body model into an FE model is called **meshing**, and it is an important step in the FE workflow. It enables the replacement of a complex engineering problem by many simpler bonded problems that have to be solved simultaneously. The software formulates a number of algebraic equations on its own, uniting them in one sparse matrix equation through the connections among the elements, the material properties of the body, the restraints and the loads. The solution of the generated matrix equation governs the behaviour of each FE and consequently relates to the entire body. The final results provide different data for the stress, displacement, strain, temperature, velocity, acceleration, etc. at each separate FE. Therefore, the accuracy of the mesh strongly affects the accuracy of the final solution.

Yet, there is no need for the user to be aware of all mathematical details that form the core of the FEM to successfully reach the correct solution. It is enough for him to be acquainted with some of the basics of FE techniques and their application through a certain program.

1.3 BASIC STEPS OF ALL ENGINEERING SOFTWARE, BASED ON FEM

There are a lot of engineering programs that use the FEM to do structural analysis. Some of them are intended to perform different specific analyses and are used in the industry and in science, while others are of a more general level and can be used even by undergraduate students. But all of these software packages have something in common the workflow and the basic stages that have to be performed. The FE software has three main stages that have to be passed through. These are the **preprocessor**, the **processor** and the **postprocessor**. No matter what they are called exactly, their functions within the programs are equivalent.

The user has to create the solid geometry of the body, assign the material properties, impose the displacement or contact boundary conditions and apply external forces in the **preprocessor**. At that level, the knowledge of functioning of the physical models is crucial for finding an accurate solution of the defined engineering problem. That knowledge complemented by a thorough understanding of the logic and the development of FE models leads directly to the final result. The user has to involve his or her

2

entire experience to successfully combine the knowledge of the operation of the physical model to the specifics and perquisites of the finite element model.

The **processor** transforms the development of the preprocessor solid body model into an FE model. Here the software performs meshing (generation of FE mesh) and runs the solution. Its interaction with the user is minimal. In fact, the software generates the core mathematical equations of the FEM almost independently and solves them. It generates all matrixes and arrays regarding the set geometry, material properties, boundary and load conditions, etc. The user can only choose, depending on the program, the size and the type of FE, the type of the used solver and some additional options. However, as a whole, his or her role is passive compared to the active participation in the first (preprocessor) stage. The outcome of the processor is a large dataset, which is systematised by the postprocessor.

The **postprocessor** produces visually or numerically all results. Thus, the user can easily systematise and analyse the data. He or she can verify and modify the model or make some improvements if necessary.

1.4 SW SIMULATION AS A PACKAGE FOR FEA

SW Simulation is integrated in some of the SW products, for example, SW Premium, SW Simulation Premium or SW Simulation Professional, enabling the development of an FEA. One of its main advantages is the close interaction between the CAD (geometrical) model and the FE one. In fact, this software is among the best examples of engineering products for CAD/FEA and design. All changes made in the geometry of the studied object are automatically transferred into the FE model, and the software reports that. All performed studies can be saved, duplicated, renamed, etc. They are organised in a tree structure, which can easily be modified.

Another plus of SW Simulation is the existence of **Simulation Advisor** (⚡). It leads the user through the analysis workflow to achieve the final result. It is recommended to be used by users who do not have enough experience either with the method or with the software.

Additionally, there are some more 'advisors,' such as Study Advisor, Bodies and Materials Advisor, Interactions Advisor, Mesh and Run Advisor and Results Advisor, which can be activated at different stages of the analysis.

Through his or her work, the user can be connected to a large database with online resources by activating the **Analysis Research** icon (🔍). He or she can **Request License Online** and can be linked to **Simulation Subscription Service** (📇). Even more, the user has a link to the **SolidWorks Simulation Web Site** (🌐), where he or she can exchange ideas with other members of **SolidWorks Simulation Community Groups** (📋) or download some files from **SolidWorks Simulation Subscription Support – Download** (📥).

Different types of analyses can be done using SW Simulation. They include static (or Stress) studies (🔧); frequency studies (🔧); buckling studies (🔧); thermal studies (🔧); drop test studies (🔧); fatigue studies (🔧); nonlinear studies, including nonlinear static study (📐) and nonlinear dynamic studies (📐); linear dynamic studies, including modal time history studies (〰), harmonic studies (〰), random vibration studies (〰) and response spectrum studies (📈); and pressure vessel design studies (🔧).

In this book, we will explain how static studies of simple bodies to more complex structures can be done.

CHAPTER 2

DEVELOPMENT OF A FINITE ELEMENT MODEL OF A BODY (PRE-PROCESSOR STAGE)

2.1 *DESCRIPTION OF FUNCTIONS OF PHYSICAL MODEL*

We will begin our introduction to SW simulation with a static analysis of a chisel (Figure 2.1).

First, we have to clarify our idea about what chisel is, where it is used and how it works. After that, we continue with the development of the CAD (geometrical) model and its transformation into a finite element (FE) model.

It must be acknowledged that the answers to the previous questions are crystal clear, and because of that, we start the introduction with this cutting tool, which is commonly widespread and familiar to everybody. However, as understanding the operation of a chisel is of significant importance for the development of a correctly

Figure 2.1
CAD model of a chisel, developed in SolidWorks.

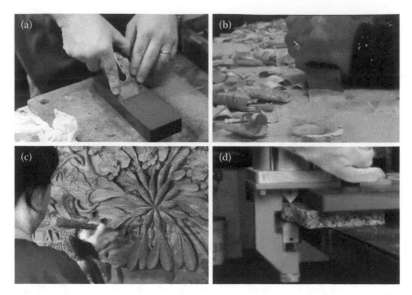

Figure 2.2
How to use a chisel. (a) Guitar building (hand production of musical instruments); (b) wood-working chisel (furniture industry); (c) wood carving; (d) chisel machine "Jaws by Monolit" (stone working industry). (Available at http://www.youtube.com.)

restrained and loaded FE model, which itself leads to accurate results, we will provide some examples of how chisels can be used. Chisels are an important tool in the wood-carving and woodworking industry, stonework, art design, etc. (Figure 2.2).

Generally, chisels are made of alloy steel. They are fixed at the root and loaded at their opposite sharp edge (the cutting edge). These two basic features of our prototype will help us later in defining restraints and loads.

We studied and clarified how the physical model operates, regarding the materials and restraints.

We know how the studied object functions. We have an idea about its geometry; therefore, we can start the development of the CAD model of the chisel.

2.2 DEVELOPMENT OF THE GEOMETRICAL MODEL IN SolidWorks

Having understood how our prototype functions, we must proceed to the next step – development of a CAD model. We can develop a CAD model through whatever software we are accustomed to and then export it in SolidWorks using one of the interchangeable formats such as IGES (*.igs, *.iges), STEP (*.step, *.spt), CATIA Graphics (*.cgr), Inventor (*.ipt, *.iam), and Solid Edge (*.par, *.psm, *.asm).

A brief instruction on how to model a chisel is provided in the following.

1. Starting a new file of type *.sldprt:

File → New () → New SolidWorks Document → a 3D representation of a single document (   )

The **SolidWorks** working environment is activated. It includes the **Menu bar, Graphics area, SolidWorks Resources** and **Status bar** (Figure 2.3).

There are two groups of commands on the **Menu bar**. The command line menu is visible when the cursor is placed over the bar or the user has clicked the SolidWorks logo. The bar can be kept visible if the user pins it using the drawing pin icon   at the right end of the menu bar. The second group of commands includes the icons of the most commonly used commands. It is always visible, and when the two bands are kept visible, it is situated on the left side of the **Menu bar (Figure 2.4)**.

2. Saving the going-to-be-developed CAD model:

File → Save as (   ) → Browsing to displace the file in the working directory → naming the file (Chisel) → Save

The file will be named Chisel to remind us of the prototype. From now onwards, the software will save all geometry data to the file *Chisel.sldprt*. Every time we want to save our model, we can use the **Save icon** (   ) on the **Menu bar**. We can reload the model through the path

File → Open (   ) → Pick the file (Chisel. sldprt) → Open

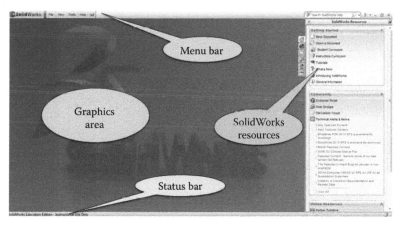

Figure 2.3
SolidWorks working environment.

Figure 2.4
SolidWorks Menu bar.

3. Setting the unit system. It will be SI system: millimetre gram second. We will follow the path

Tools → Options → Document Properties → Units → Unit System (check MMGS) → OK

Pick the command **Tools** from the **Menu bar** (Figure 2.4, Command line menu); pick **Options** (⊞) from the pop-down menu; click the **Document Properties** tab of the opened **System Options – General** window; select **Units** from the properties tree; check **MMGS**; and finally click the **OK** button to keep the introduced settings.

4. Drawing a sketch of a square with an edge of 60 mm in the **Front** plane (Figure 2.5f):

To do so, at first, we have to choose the drawing plane:

Sketch tool → Sketch → Front Plane

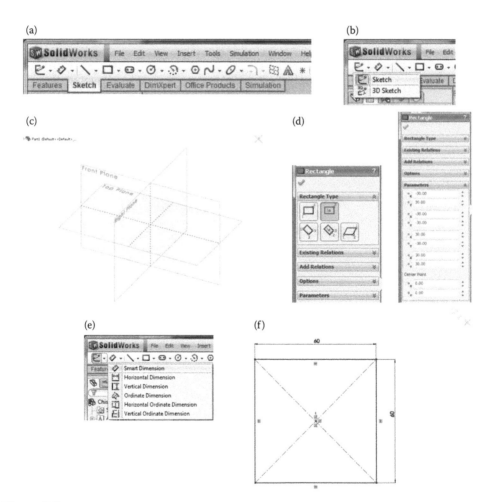

Figure 2.5
Drawing a rectangle. (a) Sketch tool bar; (b) sketch command; (c) trimetric view of initial drawing planes; (d) Rectangle property manager; (e) Smart Dimension tool; (f) drawn and dimensioned square.

We pick the **Sketch tool bar** (Figure 2.5a) and select the **Sketch** command (⬚, Figure 2.5b). The software waits until we pick a drawing plane by clicking on it at the **Graphic Area**. We pick **Front Plane** (Figure 2.5c) for our first **Sketch1**. We sketch the square through the **Center rectangle option** (⬚) of the **Rectangle** property manager (Figure 2.5d):

Sketch tool → Sketch → Rectangle → Center rectangle option (⬚) → OK

Then, we introduce the rectangle dimensions through the **Smart Dimension** tool (⬚, Figure 2.5e). For analysis purposes, it will be OK if we set the **Tolerance/Precision** to zero.

5. Defining a new plane **Plane1**, parallel to the front one at a distance of 25 mm:

Feature tool → Reference Geometry → Plane (⬚) → OK

We pick the **Feature tool bar** (Figure 2.6a) and select the **Reference Geometry** command (⬚, Figure 2.6b). From the pop-down menu, pick **Plane** (⬚) and input the plane features in the **Plane** property manager – a plane parallel to the **Front** plane at a distance of 25 mm (Figure 2.6c). The newly defined **Plane1** is shown in Figure 2.6d.

6. Drawing a circle of a diameter of 50 mm (Figure 2.7):

Sketch → Circle → Center Circle (⬚) → OK

This circle and all onward sketched circles and the square are concentric.

7. Defining **Plane2**, which is parallel to the first two and at a distance of 25 mm from **Plane1**. Sketching a second circle with a diameter of 80 mm (Figure 2.8):

Feature tool → Reference Geometry → Plane (⬚) → OK (Figure 2.8a and b)

Sketch → Circle → Center Circle (⬚) → OK (Figure 2.8c and d)

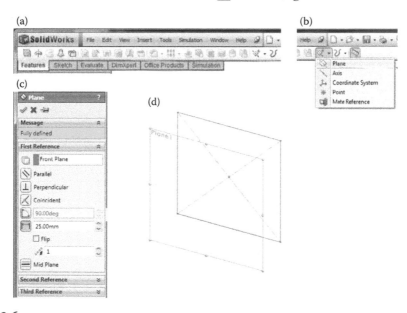

Figure 2.6
Definition of Plane1. (a) Feature tool bar; (b) Reference Geometry pop-down menu; (c) Plane property manager; (d) newly defined Plane1.

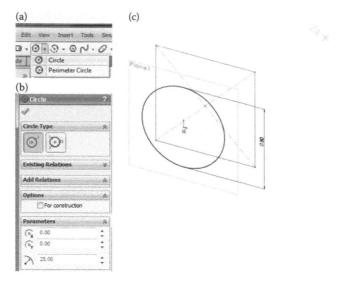

Figure 2.7
Sketching the circle in Plane1. (a) Circle pop-down menu; (b) Circle property manager; (c) Circle with a diameter of 50 mm in Plane1.

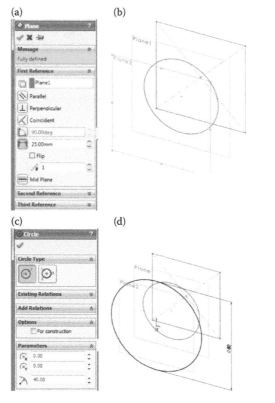

Figure 2.8
Sketching the circle in Plane2. (a) Options of Plane property manager when Plane2 is defined; (b) defined Plane2; (c) Circle property manager at drawing the circle from stage 7; (d) drawn circle from stage 7.

8. Defining **Plane3,** which is parallel to the rest of the planes and lies at a distance of 40 mm from Plane2. Sketching the third circle, with a diameter of 80 mm (Figure 2.9):

Feature tool → Reference Geometry → Plane (⊠) → OK (Figure 2.9a and b)

Sketch → Circle → Center Circle (⊙) → OK (Figure 2.9c and d)

9. Using the Loft feature, we create the root of the chisel (Figure 2.10):

Feature tool → Lofted Boss/Base (🗗) → OK (Figure 2.10a)

Right clicking in the blue **Profiles** sub-window of the **Loft** property manager (🗗) opens the pop-up menu, shown in Figure 2.10b. We pick the **SelectionManager** to help us in easier selection of the lofted contours. We then push the **Group Selection** button (🗒). Then we select all lines that outline the square (Figure 2.10c) and click the **OK** button of the **SelectionManager**. The signature of the contour is displayed in the blue window. Then we consequently select all circles and confirm each choice by clicking **OK** after each selection. The input properties of the **Loft** property manager are given in Figure 2.10d. Figure 2.10e shows the **Graphic area** view during the introduction of all contours in the **Profiles** sub-window. The green spheres and the dash line connecting them mark the guiding line of the loft. You can try to modify it by simply dragging the green spheres along the profiles. After clicking **OK** (✔) at the **Loft** property manager (🗗), the software displays the lofted root (Figure 2.10f).

The second stage of CAD modelling of the chisel is the creation of its body.

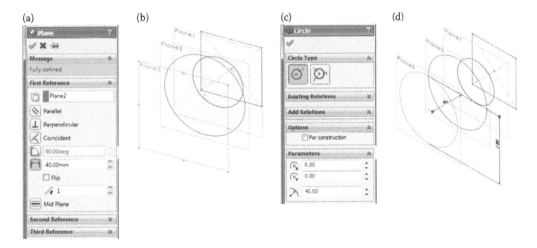

Figure 2.9
Sketching the circle in Plane3. (a) Options of Plane property manager when Plane3 is defined; (b) defined Plane3; (c) Circle property manager at drawing the circle from stage 8; and (d) drawn circle from stage 8.

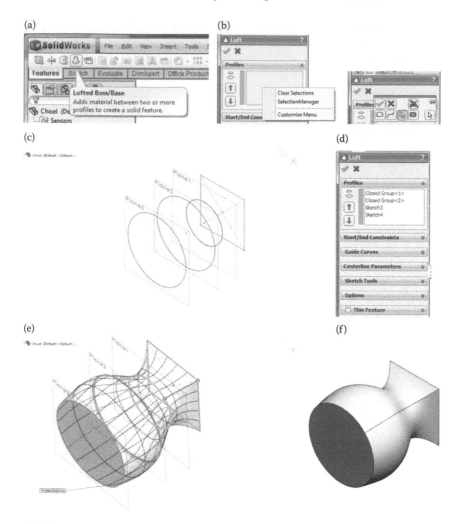

Figure 2.10
Modelling the root of the chisel. (a) Starting Loft Boss/Base command; (b) Selection manager; (c) picking the closed Group1; (d) Loft property manager with all picked contours; (e) graphic area view after all contours are picked; (f) the geometric model of the root of the chisel.

10. Definition of a new plane (**Plane4**) on the opposite side of the **Front** Plane at a distance of 200 mm and sketching there a rectangle sized 5/150 mm (Figure 2.11):

Feature tool → Reference Geometry → Plane (⊗) → OK (Figure 2.11a, b and c)

Sketch tool → Sketch → Rectangle → Center rectangle option (▱) → OK (Figure 2.11d and e)

11. Lofting the body of the chisel (Figure 2.12):

Feature tool → Lofted Boss/Base (⌀) → OK

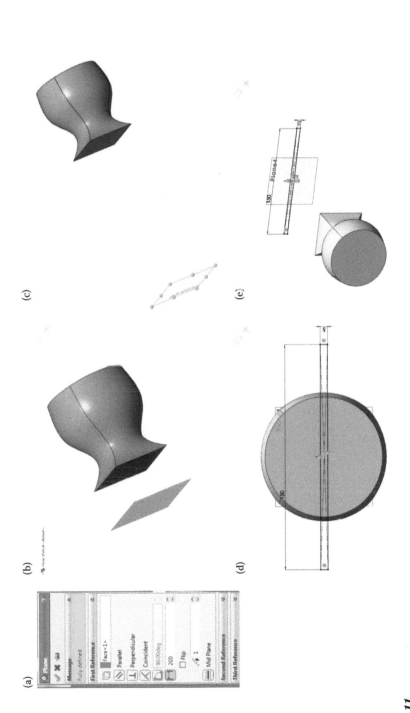

Figure 2.11

Drawing the sketch, outlining the cutting edge of the chisel. (a) Plane property manager; (b) graphic area view at defining Plane4; (c) graphic area view of the defined Plane4; (d) sketched rectangle, in which the geometric centre is collinear with the geometric centres of the circles (front view); (e) sketched rectangle, which geometric centre is collinear with the geometric centres of the circles (Dimetric view).

(a) (b)

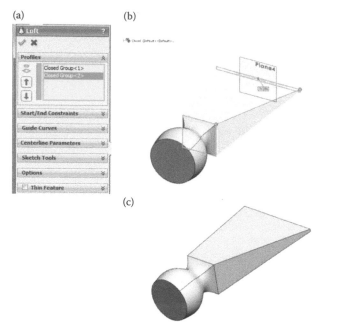

(c)

Figure 2.12
Lofting body of the chisel. (a) Loft property manager; (b) graphic area view of the lofted body of the chisel; (c) trimetric view of the chisel.

Select the **Closed Groups** (Figure 2.12a) using the **SelectionManager**, particularly the **Group Selection** button (🖼), to enable easier selection of the lofted contours. Then select all lines that outline the square and the rectangle (Figure 2.12b) to establish the two **Closed Groups**. Click the **OK** button of the **Loft** property manager to view the ready CAD model (Figures 2.12c and 2.13).

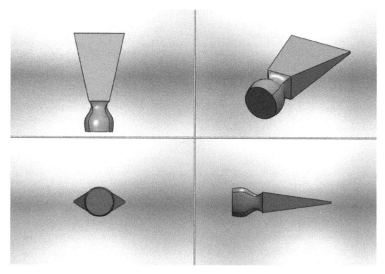

Figure 2.13
Different views of the ready chisel.

We remembered how a CAD model of a simple part can be drawn. We remind how to start developing a model, how to define the **Unit** system and how to use **Sketch** and **Feature** tools.

During that section, we are reminded how to

- Start developing a CAD model in SolidWorks
- Set the Unit system
- Sketch simple figures, such as circles and rectangles
- Define new planes
- How to feature sketches or contours using Loft Boss/base command

2.3 SOME MORE PERQUISITE KNOWLEDGE BEFORE DEVELOPMENT OF SW SIMULATION MODEL

2.3.1 Main Features of Linear Static Analysis

Finally, we have an idea about the object of our analysis; in fact, we even have the CAD model of our prototype and it seems that we are ready to start. But before proceeding with the analysis, we have to answer one more question: What is the static analysis?

There are several types of analysis that can be made through **SW Simulation**. Static analysis is one of them. It calculates the displacements, strains and stresses in a body or in a structure under the effect of applied external loads (forces, torques, temperatures, gravity, etc.) and with respect to the predefined materials and restraints (fixtures and connections). All of us know that when a body is loaded, it deforms. The effect spreads throughout the whole body. It induces changes in inner forces and reactions and renders the body into something new and totally different from the initial one state of equilibrium. We can make either a linear static analysis or a non-linear one.

This course will teach you how to make a linear static analysis ($\mathbf{q}^{*}$). Our introduction to linear static analysis will start with the analysis of the chisel. When performing a linear static analysis ($\mathbf{q}^{*}$), we have to keep in mind the assumptions about the following:

- **Static loading**. This means that all loads are applied slowly and gradually, and when they reach their maximal values, they remain constant. To be more precise, we have to explain that slowly loading means that the time interval for which the load increases its value is larger than one-third of the period of the fundamental frequency of the body.
- **Linearity assumption**. This means that the relationship between the loads and the responses is linear, that is, if we double the values of all loads, the responses (stress, displacement, strain, reactions, etc.) will also double (Figure 2.14a). To validate that assumption, we have to be certain that
  - The Hooke's law is applicable and the stress is proportional to the strain (Figure 2.14b).
  - All material properties, such as **Young's modulus** and **Poisson's ratio**, remain constant during the analysis.
  - The restraints and the loads do not change during the deformation.

The final state of the body does not depend on the consequence of applying the loads.

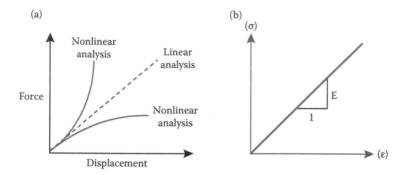

Figure 2.14
Functions, describing linearity assumption [SW Simulation On-line Help]. (a) Comparison between linear and nonlinear "force-displacement" function; (b) Hook's "strain–stress" diagram.

2.3.2 Starting SolidWorks Simulation

We are now ready to start our first analysis.

- We know and understand how our object operates; hence, we have enough knowledge to discuss the material, the restraints and the external loads.
- We have a ready CAD model.
- We have a brief idea of the type of analysis we are going to do and the kind of results expected to be received.

To start the SW Simulation tool, we have to pass through some stages, which are described further.

2.3.2.1 Activate SW Simulation Toolbox To activate the **SW Simulation toolbox** (Figure 2.15), we have to follow the path

Tools → Add-Ins → SW Simulation → Close

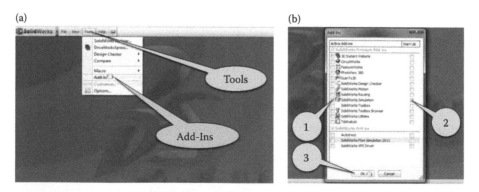

Figure 2.15
Commands for activation of SW Simulation. (a) Opening the Add-Ins window; (b) activating SW Simulation toolbox.

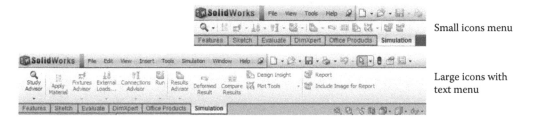

Small icons menu

Large icons with text menu

Figure 2.16
SW Simulation command bar.

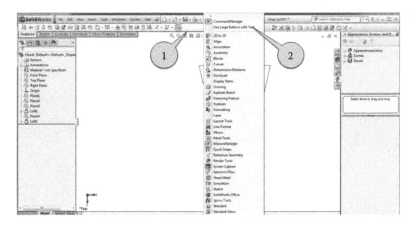

Figure 2.17
Pop-down menu helping to customize the displayed toolbars.

When the **Add-Ins** window is opened, you must select the pointed buttons (Figure 2.15b) to

- Activate the SW Simulation toolbox
- Keep it active when SW is started the next time
- Confirm the commands

When the SW Simulation is activated, a new command bar appears below the **Menu bar** (Figure 2.16). Most of the icons are inactive and grey as no analysis is defined yet.

The user can choose to use either large icons or small icons (Figure 2.16). Large icons can be used by right clicking on the command bar and checking the **Use Large Buttons with Text** line on the pop-down menu (Figure 2.17). For beginners, the menu with large icons with text is recommended.

2.3.2.2   Open the CAD Model   To start the analysis, a CAD model must be opened (Figure 2.18):

File → Open → browse for the file *.sldprt (Chisel.sldprt)

When the CAD model (henceforth, we will call it simply a model) is opened, the working area will look as in Figure 2.19, and we are now ready to start our first FE analysis through SW Simulation.

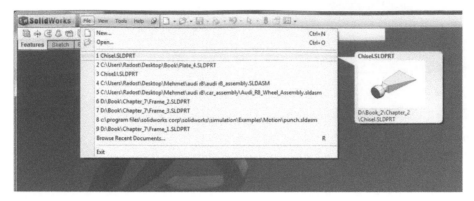

Figure 2.18
How to open an existing CAD model.

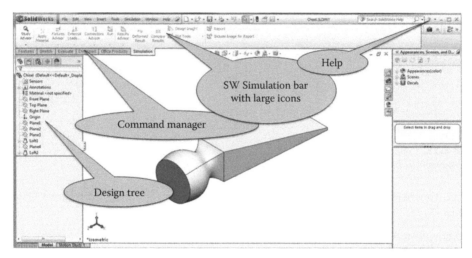

Figure 2.19
CAD model of the chisel – working environment.

2.3.2.3   Getting Access to Help Files   At each stage of the analysis, even at the very beginning, you are able to ask for **Help**. As the **SW Simulation** tool is activated and the model is loaded, you have access to some other types of **Help** (Figure 2.20a), particularly focused on simulations. They involve

- **SW Simulation Help Topics**, with some theories on the method and explanations of the functions and options of the commands (Figure 2.20b)
- **SW Simulation Tutorials**, with some examples where each step is carefully explained (Figure 2.20c)
- **SW Simulation Validation**, with some verification problems and National Agency for Finite Element Methods and Standards (NAFEMS) benchmarks (Figure 2.20d)

To start the analysis, you can either click on the **Study Advisor** icon (Figure 2.21) or even simpler on the icon (🔍). As a result, the **Simulation Advisor** (🏁) is activated.

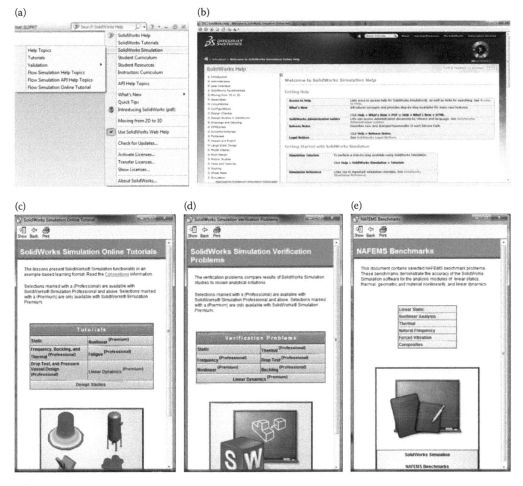

Figure 2.20

SW Simulation help. (a) SW Help; (b) SW Simulation online help topics; (c) SW Simulation tutorials; (d) SW Simulation verification problems; (e) SW Simulation NAFEMS benchmarks.

This is a window on the right side of the working area (Figure 2.21b). It guides the user through the analysis process, and if you follow the instructions and answer the questions in the window, you will be able to perform your analysis successfully. Meanwhile you will be consulted by all five advisors of the team of the **Simulation Advisor**, which are

- Study Advisor
- Bodies and Materials Advisor
- Interactions Advisor
- Mesh and Run Advisor
- Results Advisor

Additionally, some individual advisors will guide you (Figure 2.21b and c). They can be accessed either through the icons that are situated on the **SW Simulation** bar or through the icons that appear in the simulation model on the left side of the working area (Figure 2.21a). You can see that all icons in the **SW Simulation** bar are highlighted now (Figure 2.22).

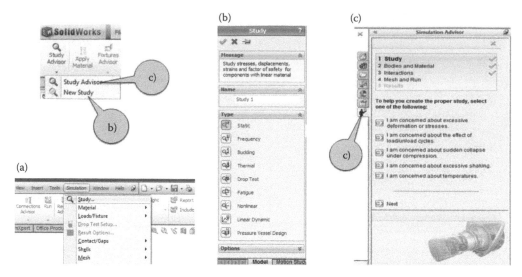

Figure 2.21
*Start of a new study and of the individual advisors. (a) Start of a new study; (b) Study advisor;
(c) Simulation advisor.*

Figure 2.22
SW Simulation individual advisors. (a) At the command bar; (b) at the analysis tree.

The **Study advisor** and the **Simulation advisor** (Figure 2.21b and c) enable the user to apply the material (⫶), to define fixtures (⊿) and external loads (⊞), to add connections (⬚), to run the analysis (⬚) and finally to systematise the results (⬚).

The next step is to choose the type of the analysis and to start it. This can be done by using the **Study** property manager (Figure 2.23). The first stage is to introduce the name of the analysis (1). After that, we choose the type of the analysis (2a), and a brief description of the chosen type of the analysis is immediately displayed by the program in the yellow window in sub-window **Messages** (2b); finally, we can click either the **OK** (3a, ✔) or **Cancel** (3b, ✖) icon to preserve or to reject the input properties.

The choice of analysis is significant for "getting the right answers" by the software. **SW Simulation** performs

- **Static** *(or Stress)* **study** (⬚). It helps to avoid failure due to high stress. Static studies calculate displacements, reaction forces, strains, stresses and factor of safety distribution.
- **Frequency study** (⬚). It helps to avoid failure due to excessive stresses caused by resonance. Frequency studies calculate the natural frequencies and associated mode shapes. It provides information for solving dynamic response problems.

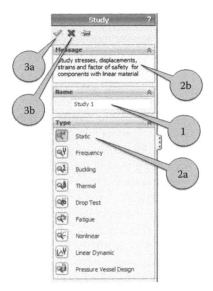

Figure 2.23
How to start a new analysis.

- **Buckling studies** (⬚). They help to avoid failure due to buckling. It occurs when slender structures are subjected to axial loads and sudden large displacements arise. Usually only the lowest buckling load is of interest, and it is lower than those required to cause material failure.
- **Thermal studies** (⬚). They help to avoid undesirable thermal conditions, like overheating or freezing. Thermal studies calculate temperatures, temperature gradients and heat flow based on heat generation, conduction, convection and radiation conditions.
- **Drop test studies** (⬚). They help to simulate the impact of the model with a rigid planar surface. Drop test studies calculate different parameters of the process to evaluate the effect of a dropping body on a rigid floor.
- **Fatigue studies** (⬚). They help to avoid weakening of the object due to loading and unloading over time even when the induced stresses are considerably less than the allowable stress limits. Fatigue studies evaluate the consumed life of an object, in relation to fatigue events, and based on fatigue calculations on stress intensity, von Mises stresses or maximum principal alternating stresses.
- **Nonlinear studies**, including **nonlinear static study** (⬚) and **nonlinear dynamic study** (⬚). They are used to solve problems with nonlinearity caused by material behaviour, large displacements and contact conditions.
- **Linear dynamic studies**, including **modal time history studies** (⬚), **harmonic studies** (⬚), **random vibration studies** (⬚) and **response spectrum studies** (⬚). They use natural frequencies and mode shapes to evaluate the response of structures to dynamic loading environments.
- **Pressure vessel design studies** (⬚). They combine algebraically using a linear combination or the square root of the sum of squares (SRSS) and the results of static studies under different sets of loads.

Within this course, we are going to perform **Static analyses** (⚙) of bodies and structures. Therefore, we click the corresponding icon (⚙) and introduce the name of the study, which by default is **Study 1** (1, Figure 2.23) but can be changed to any name that describes the analysis better. It is very important not to forget to click **OK** (✔) or **Cancel** (✖). This automatically closes the **Study** property manager (Figure 2.23).

The next step is to introduce the properties of the started study (Figure 2.24). We can access the **Study panel** either through the **Study Advisor menu** (Figure 2.24a) or through the **Static analysis tree** by right clicking on the name of the analysis (Figure 2.24b). If we use the **Study Advisor menu**, we must click the **Study Properties** line, or if we open the **Static analysis tree** pop-down menu, we must click **Properties**.

The newly opened **Study properties** dialog window has four different panels, which allow the user to introduce different characteristics of the analysis by choosing options and answering to a set of questions. The **Study properties** dialog window involves four different sub-windows (Figure 2.25).

The first sub-window to be accessed is **Study options** (Figure 2.25a). All properties of the on-going analysis can be introduced through this window. There are some features that the software has already selected. In fact, they are introduced by default; however, the user can change them if necessary. For newly accustomed users, this action is not recommended.

The first tab enables the definition of **Gap/Contact options**:

- **Include global friction**. This controls the inclusion of the effect of friction for global contact conditions. The software calculates static friction forces by multiplying the normal forces generated at the contacting locations by the friction coefficient, which is introduced through the window on the left and has a value in the range of 0 to 1.0 (0.05 by default).
- **Ignore clearance for surface contact**. This enables considering the contact conditions regardless of the initial distance between user-defined face pairs.
- **Improve accuracy for no penetration contacting surfaces (slower)**. This results in continuous and more accurate stresses in regions with definitions of no penetration contact.

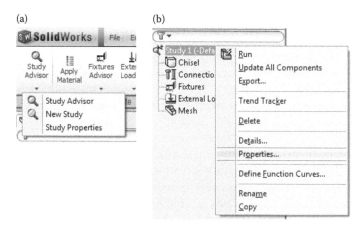

Figure 2.24
Starting the Study properties dialog window. (a) Study Advisor menu; (b) Static analysis tree.

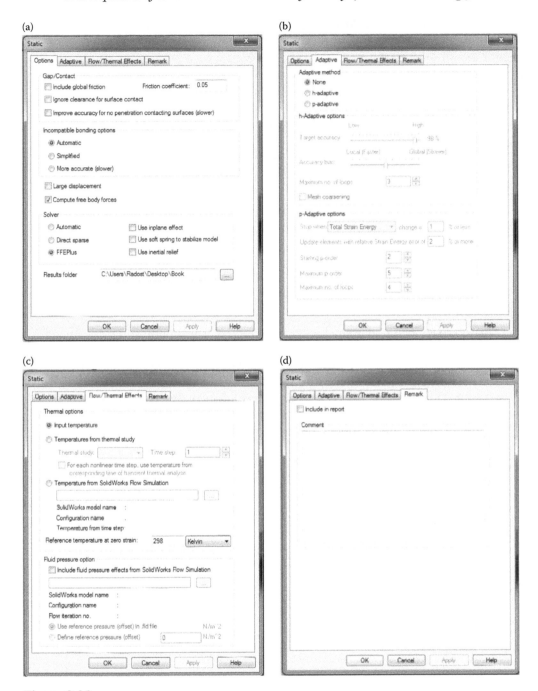

Figure 2.25
Introducing of the Study properties. (a) Study options sub-window; (b) Adaptive options sub-window; (c) Flow/Thermal effects sub-window; (d) Remarks sub-window.

The second tab on the **Study options** sub-window enables the definition of **Incompatible bonding options**. The user has to choose among three possible approaches of calculation:

- **Automatic**. The default bonding contact is surface to surface, but the solver can switch automatically to node-to-surface bonding contact to accelerate the calculations.
- **Simplified**. This is recommended when solving models with extensive contact surfaces.
- **More accurate (slower)**. The surface-to-surface bonded contact is applied through the entire calculation.

The next step is to choose between

- **Large displacement**: when the program applies the loads gradually and uniformly in steps up to their full values performing contact iterations at every step.
- **Compute free body forces**: the program keeps the force balanced at every node at each node of the FE, including external loads, restraint or contact reactions, etc. This option is chosen by default.

The third tab is the **Solver** tab. In this tab, the user can choose among the following options:

- **Automatic**: the program makes the choice itself. This option is recommended for linear static studies, such as all examples included in the course.
- **Direct Sparse**: it is recommended in cases of multiarea contact problems. It provides satisfying efficiency in speed and memory usage for small problems (up to 25,000 DOFs). When you have enough memory on a computer, the Direct Sparse solver is quicker than the FFEPlus one. Additionally, it is recommended when there are materials varying in wide range material properties, particularly moduli of elasticity.
- **FFEPlus (Fourier Finite-Element Plus)**: it is strongly recommended for solving large problems (over 300,000 DOFs). It is faster than the Direct Sparse solver as the problem gets larger (Figure 2.26).

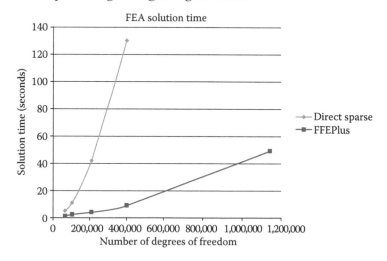

Figure 2.26
FEA solution time versus number of DOFs. (Available at http://www.javelin-tech.com/blog/2013/01 /which-solver-ffeplus-vs-direct-sparse-part-1/.)

If you choose one of the two solvers, you can control the use of in-plane effect, the use of soft spring to stabilise the model and the use of inertial relief. The **Use soft spring to stabilize model** is recommended if the design is unstable, and it is necessary that more restraints be activated to prevent the motion. The flag of that option should not be active by default. The software applies inertial forces to counteract unbalanced external loading when the **Use inertial relief** flag is on. While solving structural problems, this option enables finding the correct solution even if there are not enough restraints and the soft spring option is off.

Finally the user can specify the directory where the simulation results should be stored – **Results folder**.

After choosing all options regarding the purpose of the analysis or keeping them as they are by default, you must click on the **OK button** to save your choice.

The second sub-window sets the adaptive options of the static study and can be accessed through clicking the **Adaptive** tab (Figure 2.25b). Two main adaptive methods, based on error estimation, are used by **SW Simulation** – the **p-method**, which does not change the mesh but increases the FE order to improve the results, and the **h-method**, which refines the mesh but keeps the element order. The **h-method** is recommended for bodies with complex geometry and loading, including sharp corners and concentrated loading. It uses smaller elements in regions with high errors and automatically refines the mesh. The **p-method** increases the FE order, which means an increase in the order of the polynomials used to approximate the displacements. As this is not effective to be done for all elements, the software selects the regions, that is, the selective adaptive p-method is adopted. In this release, the **p-method** does not work with shells and with non-uniform pressure, non-uniform forces or multiple pressures defined on a face.

In the first tab of the **Adaptive** sub-window, the user can choose among the use of no adaptive methods (None button, which is checked by default) and the two adaptive methods.

The **h-Adaptive options** are

- **Target accuracy**: the higher the percentage is, the more accurate the final stress results are, yet the calculation level increases.
- **Accuracy bias**: if the slider is closer to **Local**, the peak stresses are in the focus of the solver, and if the slider is closer to **Global**, the software focuses on the overall stress accuracy.
- **Maximum no. of loops**: sets the maximum number of loops allowed but no more than 5. It is 3 by default.
- **Mesh coarsening**: if the flag is on, the software coarsens the mesh in regions with low error during the adaptive loops.

The software stops calculations based on either of the above defined limits.

The **p-Adaptive options** are as follows:

- **Stop when**: sets the global criterion for convergence and termination of the loops; can be **Total strain energy** (the sum of the strain energy of all elements); **RMS von Mises Stress** (the root mean square value of the nodal von Mises stresses); or **RMS Res. Displacement** (the root mean square value of the nodal resultant displacements). The **maximum allowable relative change** is set as a percentage.
- **Update elements with relative Strain Energy error xx% or more**: if none of the two stopping criteria defined above are met, the program increases the polynomial order of the elements according to this criterion.

- **Starting p-order**: sets the order to be used for the first loop and varies between 2 and 5.
- **Maximum p-order**: sets the highest p-order to be used. The limit is 5.
- **Maximum no. of loops**: sets the maximum number of loops allowed in the analysis. The limit is 4.

The program stops the loops when one of the above conditions is met.

The third tab of the **Study** property manager is **Flow/Thermal Effects** (Figure 2.25c). If there are no redundant restraints at the body, the changes in temperature cause no additional stresses, but if the body is prevented from free elongation or contraction, the so-called temperature stresses are induced. Consequently, the thermal effects as a consequence of temperature variations have to be studied and added to the stress impact of the loads for all structures with redundant restraints. As this is not our case, we will not discuss this in detail now. It is enough to know that introducing the thermal effects to our analysis can be done through the following options – **Input Temperature; Temperatures from Thermal Study and Temperatures from Flow Simulation**. The same is the situation with the introduction of the **Fluid pressure option**, where the pressure distribution function is input from a **FlowSimulation** results file.

All remarks to the study can be introduced through the **Remark tab** (Figure 2.25d) and will optionally be included in the final report. The desired text is typed in the **Comment** window and is confirmed by clicking the **OK** button.

For the performed static study, it is accepted that all Static analysis properties be left as they are by default, that is

- **Options**: no Gap/Contact options; Automatic incompatible bonding options; Compute free body force; Automatic solver; Results folder – the directory of the open part file.
- **Adaptive**: no adaptive method is selected.
- **Flow/thermal effects**: neither thermal nor flow effects are introduced.
- **Remark**: no remarks.

We summarised all types of the analysis that SW Simulation can perform. We discussed all analysis properties that influence the FE analysis and pointed out the advantages of the options suggested by the program.

Up to now, we learned how to

- Start SW Simulation tool
- Set the type of the analysis
- Use the built-in help
- Define properties of the static analysis

2.4   INTRODUCING THE MATERIAL OF THE BODY

2.4.1   How SW Simulation Handles Material Properties

Finally, we have an idea about the object of our analysis. After defining the properties of the analysis or leaving them as they have been defined by the software, it is time to start with the introduction of the model characteristics, particularly materials, fixtures, loads and contacts. **SW Simulation** transfers them directly to the solid body model; hence, they have to be introduced in the pre-processor stage. If they need to be modified later, the software automatically applies the changes, prompts the user and re-meshes the model.

Thus, the model development continues with the definition of the materials. This can be done either by clicking on the **Apply Material** icon (≋Ξ) on the command bar or by achieving the command through the analysis tree following the path (Figure 2.27)

Body (right click) → Apply/Edit Material

As a result, a new window opens, where we can either choose a material or define a new one. The definition of a material in SW Simulation does not update automatically the materials assigned to the CAD model.

The **Material** property manager is shown in Figure 2.28.

On the left side of the **Material** property manager, the **Material Tree** is set (Figures 2.28 and 2.29). There are three basic groups of materials: **SolidWorks library materials** are split in **SolidWorks DIN Materials** and **SolidWorks Materials**, and **Custom Materials**. By choosing the last material group, the user can define a new material introducing its name and group as well as its material properties, and later on manage them into a user-defined library.

For non-experienced users, it is recommended to use a pre-defined SW material instead of defining a new one.

For static analysis, we can choose either **isotropic** or **orthotropic material**. The isotropic materials possess the same mechanical and thermal properties in all directions (say, steel), whereas the orthotropic materials demonstrate different mechanical and thermal properties in the three orthotropic directions (say, wood). The assumption of linearity is active for both types of materials.

When defining, choosing or even editing a material, it is necessary to change the data in the **Properties** dialog box (Figure 2.30). It is used to assign the physical properties of the material. The **Model Type** describes the stress–strain relation of the material, and as has been said, only **Linear Elastic Isotropic** and **Linear Elastic Orthotropic** material types are available for static analysis. The **Unit** window sets the unit system in which the values of the material properties are displayed. We shall use the SI system in this

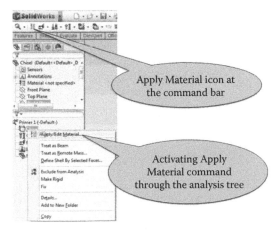

Figure 2.27
Apply Material command.

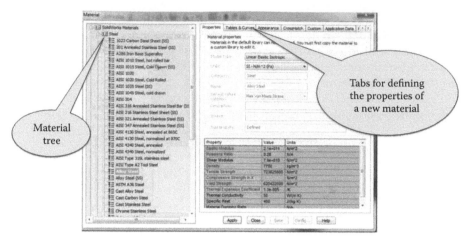

Figure 2.28
Material property manager.

Figure 2.29
Material tree.

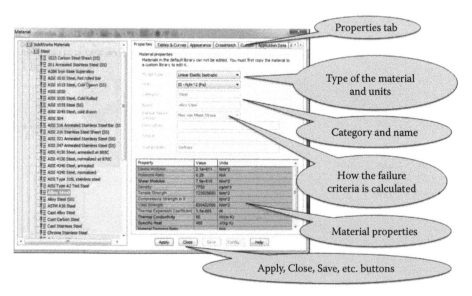

Figure 2.30
Material property manager – Property window.

course. After that, the **Category** to which the selected material belongs and its **Name** are introduced. Through the **Default failure criterion**, the user can select which failure criterion to be used when the factor of safety is calculated. In the **Description** window, a free text up to 256 characters can be typed. The **Source** window is active only when a Custom Material is used. The **Sustainability** window indicates if the material has a link to the sustainability database. The **Material Property Table** comes last. It consists of three columns – **Property**, where the material properties are listed and the list depends on the selected material type and the study type; **Value**, where the numerical value of each material property is displayed or input; and **Units** – lists the unit of each material property. Sometimes the **Temp Dependency** column can be added. Material properties, which values are mandatory for static analysis, are coloured in red (elastic modulus, Poisson's ratio, etc.); those that are optional (tensile strength, compressive strength, etc.) are blue; and those that have values that can be calculated by the program (shear modulus) or are not directly involved in the calculations remain black. The mandatory values and the colouring of the material properties depend on the type of the analysis. The basic material properties are

- **Elastic modulus** (also called Young's modulus and modulus of elongation): for a linear elastic material, the elastic modulus is equal to the ratio between the stress and the associated strain in that direction. It is obligatory for strain–stress calculations.
- **Poisson's ratio**: defines the relations between the longitudinal and the lateral strain. For isotropic materials, it is equal in all directions. Poisson's ratio is a dimensionless quantity.
- **Shear modulus** (also called modulus of rigidity): the ratio between the shearing stress in a plane divided by the associated shearing strain. If a linear isotropic material is chosen, the software can calculate the shear modulus using the values of the modulus of elasticity and Poisson's ratio. Hence, the value of modulus of rigidity is of no importance for the analysis.
- **Tensile strength.**
- **Compressive strength.**
- **Yield strength.**

These last three material properties are not directly involved in the strain–stress calculations. They are used to calculate the factor of safety or failure criteria. Consequently, depending on the chosen formulae for failure assessment, one of them is coloured in red and the program demands a value.

- **Density**: equals the mass per unit volume. Its colour is red for its value is mandatory when gravity or centrifugal loads are defined.
- **Coefficient of thermal expansion**: defined as change in normal strain per unit temperature and is used in thermal analyses.
- **Thermal conductivity**: the rate of heat transfer through a unit thickness of the material per unit temperature difference.
- **Specific heat**: the quantity of heat needed to raise the temperature of a unit mass of the material by one degree of temperature.

Both previous properties are used in thermal analyses also.

- **Material damping ratio**: used in dynamic analyses; therefore, its value is optional for static analyses.

There are more tabs at the top right of the **Material** property manager.

- **Tables and curves** tab: used to define temperature-dependent curves, that is, the module of elasticity versus temperature, Poisson's ratio versus temperature, density versus temperature, yield strength versus temperature, etc. They are introduced either by direct input of the data values or by import of the **Curve Data Points File** (*.dat). The function can be visualised.
- **Appearance** tab: used to associate a new colour or texture with the selected material.
- **Crosshatch** tab: used to select the crosshatch pattern associated with the display of the material in section views of drawing documents.
- **Custom** tab: used to add non-standard properties to the material.
- **Application Data**: used to record notes about the selected material.
- **Favorites**: used to manage the material favorites list.

After verification of the material properties, the user must click **Apply** to keep the values and **Close** to close the **Material** dialog box (Figures 2.28 and 2.30).

2.4.2 Defining the Material of the Chisel

For the performed static study, we will use **Alloy Steel**:

$$\text{SW Materials} \rightarrow \text{Steel} \rightarrow \text{Alloy Steel} \rightarrow \text{Apply} \rightarrow \text{Close}$$

This is a **Linear Elastic Isotropic** material, and all values defining its properties in **SI** units – N/mm² (MPa) – are (Figure 2.31)

- **Elastic modulus in X** – 210,000 N/mm²
- **Poisson's ratio in XY** – 0.28 N/A (dimensionless)
- **Shear modulus in XY** – 79,000 N/mm²
- **Mass density** – 7700 kg/m³
- **Tensile strength in X** – 723.83 N/mm²
- **Yield Strength** – 620.42 N/mm²

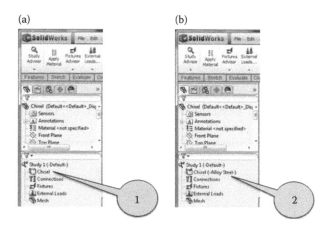

(a)   (b)

Figure 2.31
Setting the material of the chisel. (a) SW Simulation analysis tree before setting the material; (b) SW Simulation analysis tree after setting the material.

We summarised the main properties of linear isotropic materials and discussed how these properties will be involved in further calculations. We pointed out that a new custom material can be defined and added to the existing library of materials.

Up to now, we learned about:

- Existing SolidWorks libraries of materials
- Different material types and properties
- How the software grades the importance of material properties considering the type of the analysis
- The definition of a new custom material

2.5 INTRODUCING THE FIXTURES TO THE BODY

2.5.1 Different Fixtures Supported by SW Simulation

The third stage in the transformation of the CAD model into a solid body model ready for FE analysis is the inclusion of the fixtures and the definition of more restraints, if there are any (contacts, for example). The fixtures in the SW Simulation tool are fully associative and automatically adjust to every change in the geometry of the model. If a restraint section of the model is deleted or excluded from the analysis, the software immediately reports a problem.

The introduction of the fixtures can be done either through the **SW Simulation** command bar or through the **SW Simulation** analysis tree (Figure 2.22) by clicking on the icon **Fixtures** (⏚). Just as it was when we had to define the material, we can use one of the two ways to activate the **Fixtures** command (Figure 2.32).

Each rigid body can move in six independent manners in the space. Each motion of the body from one point to another can be expressed as a chain of some of these motions without having in mind their consequence. The body can make three translations parallel to the three orthogonal axes and three rotations around these axes.

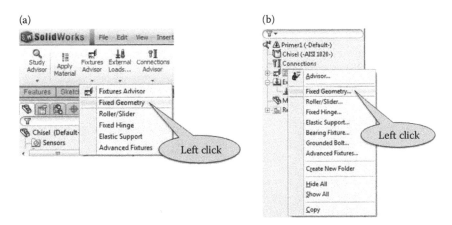

Figure 2.32
Defining the fixtures. (a) Defining fixtures through the SW Simulation command bar; (b) defining fixtures through the SW Simulation analysis tree.

Thus, it is assumed that the body has 6 degrees of freedom (DOFs). Adding a restraint or a fixture to the rigid body limiting or stopping its motion along the restricted DOF. When a simple fixture is added, it stops the translation/rotation of the point, and the number of DOFs of the body is reduced by 1. Of course, there are complex fixtures, which limit more than 1 DOF and which are even more widely spread than the simple ones. If enough DOFs are restricted, the body can stay in equilibrium in space, and if it is loaded, only displacements due to its deformations raise. If we have to solve a static problem and the number of the fixtures is equal to 6, the body is steady and the supporting loads/reactions (these are the forces that fixtures apply to the body) can be found through six independent equilibrium equations. If the number of the fixtures is smaller than 6, at least one motion of the rigid body is enabled and it functions as a mechanism. This case will not be discussed here. If the number of fixtures is greater than 6, the body is over-restrained and statically undetermined. The reactions cannot be calculated using simple equilibrium equations, yet the solution is possible to be found. In this case, the geometry and the material of the body have to be considered.

In FEM, fixtures can be applied at each node of the FE. There are two basic types of arrow used by **SW Simulation** to express the type of the fixture (Figure 2.33). The fixture in the left picture (Figure 2.33a) restrains only one translational DOF, while the fixture in the right picture ((Figure 2.33b) restricts all six possible motions. When a simple arrow visualises the fixture, it stops only the translation – one arrow, one translation is set to zero. If the fixture is visualised as an arrow with a disc at its root, it stops the translation along its body as well as the rotation around it. When there are three arrows with discs along the three orthogonal axes, all possible motions are stopped and that node is fixed.

SW Simulation uses different symbols to visualise fixtures and loads. It applies green colour to arrows that outline the disposition and the type of the fixtures. Of course, the user can redefine the colour and the size of all used arrows, depending on the model and their personal taste. This change is applicable to all symbols used by SW Simulation, its realisation is explained in the following.

The window **Symbol Settings** (Figure 2.34) is situated at the bottom of the dialog window for defining fixture, load, etc. By default, the size of the symbols is set to 100, but it can be changed easily (Figure 2.34a). If the **Show preview** is checked, the changes appear on the model automatically and are kept by clicking **OK** (✅) or **Cancel** (✖).

As has been said, the colour of the symbols is green but can be changed by clicking on the **Edit Color** button (Figure 2.34b). The **Color** property manager opens (Figure 2.35). We can choose between **Basic colors** (Figure 2.35a) and **Custom colors** palettes

(a)   (b)

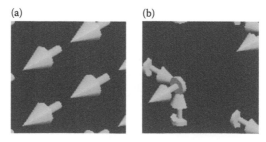

Figure 2.33
Different ways to visualise a fixture. (a) Sliding fixture; (b) steady fixture.

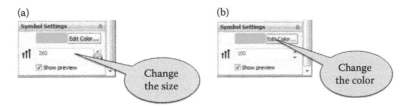

Figure 2.34
Symbol Settings window. (a) Introducing the size of the symbol; (b) introducing the colour of the symbol.

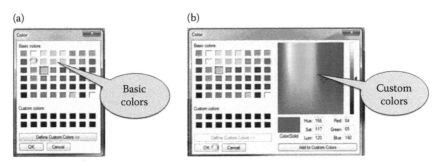

Figure 2.35
Colour window. (a) Basic colours palette; (b) custom colours palette.

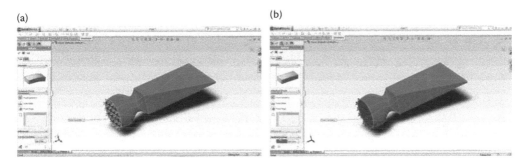

Figure 2.36
Different ways to visualise the fixture symbols. (a) Yellow fixture symbols; (b) violet fixture symbols.

(Figure 2.35b). Choosing a basic colour is enough to click on the coloured square, and after that, we click **OK**. In the presented example, the colour yellow is chosen (Figures 2.35a and 2.36a).

In Figure 2.36a, the yellow fixture symbols are shown. The new colour of the arrows is updated in the **Symbol Settings** sub-window and is seen on the model (Figure 2.36a).

Sometimes, especially in complex models with a lot of different fixtures/loads, it is recommended to use a custom colour. This is done through the **Color** property manager:

Define Custom Colors → picking the colour → Add to Custom Colors → OK

We have chosen violet as our second option (Red = 84, Green = 65, Blue = 150; Figures 2.35b and 2.36b).

Much more important for each model is the type of the fixture instead of its visual appearance. All fixtures are united in two groups: **Standard (Fixed Geometry)**, involving Fixed Geometry, Immovable Geometry, Roller/Slider and Fixed Hinge, and **Advanced**, involving Symmetry, Circular Symmetry, Use Reference Geometry, On Flat Faces, On Cylindrical Faces and On Spherical Faces. They are directly applied to the solid body model, and in the next stage of the analysis, the software transfers the fixtures to every node of each FE. The basic properties of each fixture can be summarised as follows:

- **Fixed geometry** (, Figure 2.37a). For solid bodies (3D FE) and truss joints (1D FE), this fixture sets to zero all three translations, while for shells (2D FE) and for beams (1D FE), it sets to zero all 6 DOFs. It can be applied to faces, edges, vertices or beam joints. In the **Graphic area**, it is visualised on the model with three arrows with a disk at the end ().
- **Immovable (no translation)** (, Figure 2.37b). It is reached through the **Standard (Fixed Geometry)** property manager, and it is applicable only to shells, beams and trusses. It is not accessible when the model is a solid body. It sets all translations to zero. This fixture can be applied to faces, edges, vertices or beam joints. It is visualised on the model as .
- **Roller/slider** (, Figure 2.37c). This fixture is applicable to planar faces. It sets to zero the motion in direction normal to the face and allows free motion within the plane of the face. It is visualised as on the model.
- **Fixed hinge** (, Figure 2.37d). This fixture enables the relative rotation of two cylindrical faces. It sets to zero all translations and the two rotational DOFs and frees only the rotation around the axis of the selected cylindrical face ().

The next is the group of **Advanced** fixtures.

- **Symmetry** (, Figure 2.38a). This fixture helps in reducing the model and still obtaining accurate results. The results for the "cut" part of the model are

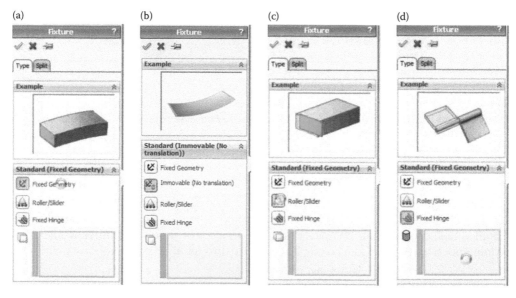

Figure 2.37
Different Standard fixtures. (a) Fixed geometry; (b) immovable (no translation); (c) roller/slider; (d) fixed hinge.

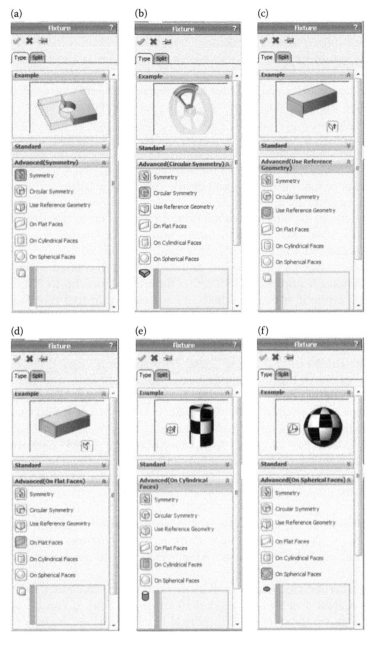

Figure 2.38
*Different Advanced fixtures. (a) Symmetry; (b) circular symmetry; (c) use reference geometry;
(d) on flat faces; (e) on cylindrical faces; (f) on spherical faces.*

deduced from the ones of the modelled part. If applied, this fixture requires
symmetrical consideration about geometry, materials and restraints, including
either loads or fixtures. It can be applied to solid bodies and to shells. When
applied to a solid body, it restrains only one translation (🝳), and when applied
to a shell, it restrains 1 translational + 2 rotational DOFs (🝳🝳). This fixture can be

applied only to faces, and this must be kept in mind when working with shells. In such cases, the symmetry fixture restrains the motion in the direction normal to the shell surface as well as rotation about the other two orthogonal axes.

- **Circular symmetry** (⌾, Figure 2.38b). This fixture is used to reduce the size of a circular model by studying a segment of it. The geometry, materials, restraints and loads are similar for all identical segments that form the entire model. The fixture is applicable to faces in solid bodies and is active only in static analysis.
- **Use reference geometry** (⌾, Figure 2.38c). This fixture uses reference geometry to apply restraints. The restrained DOFs depend on the type of the model (solid body or shell); on the reference (model face, reference plane, model edge or reference axis); and on the restraints, which can be applied to faces, edges, vertices and joints. Generally, this fixture restraints for solids – up to 3 translational DOFs (⬉); for shells and beams – up to 3 translational and 3 rotational DOFs (⬈); and for trusses – up to 3 translations.
- **On flat faces** (▱, Figure 2.38d). This fixture is applicable only to flat faces. More than one face can be selected, and each face is restrained according to its own directions. If applied to solids, this fixture restrains up to 3 translational DOFs (⬉), while if applied to shells, it can restrain up to 3 translational and 3 rotational DOFs (⬈).
- **On cylindrical faces** (▤, Figure 2.38e). This fixture is applied only to cylindrical faces and operates as the **Flat Faces** fixture.
- **On spherical faces** (◯, Figure 2.38e). This fixture is applied if the selected faces are spherical. It can be applied either to solid bodies or to shells.

A more detailed explanation about the application of the **Advanced** fixtures will be provided in Chapter 6.

Some more fixtures that can easily be accessed through the **SW Simulation** analysis tree are shown in Figure 2.32b. They are not typical fixtures. They define how the selected entity (face, edge or vertex) is connected to the ground, without need of any detailed geometry modelling. They are given as follows:

- **Elastic support** (▦, Figure 2.39a). It is used to simulate elastic foundations and shock absorbers. It is applied to faces and resists tension and compression. The definition of this connector requires data for its normal and tangential stiffness of the foundation.
- **Bearing fixture** (⬤, Figure 2.39b). It is used to simulate the interaction between a shaft and another rigid shaft or the ground. It can be applied to cylindrical faces, to concentric cylindrical faces of smaller angles of the shaft or to cylindrical shell edges. The fixture enables the adding of self-aligning features to the bearing.
- **Grounded bolt** (⬍, Figure 2.39c). It is used to connect the component (solid body or shell) to the ground. For defining the ground bolt, it is mandatory to define a reference plane. Additionally, the elastic modulus and Poisson's ratio of the bolt material and its shank diameter are introduced.

Before choosing the type of a fixture, you can choose to apply it to a section of a face instead of to a whole face. This can be done through the **Split** tab, which is situated beside the **Type** tab (Figures 2.37 to 2.39).

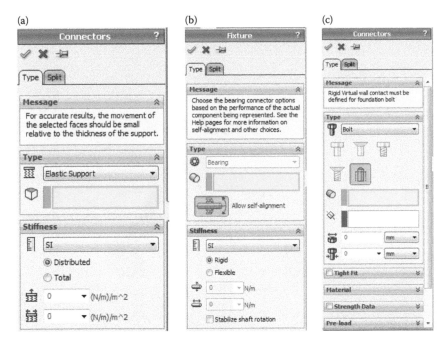

Figure 2.39
Different Connecting fixtures. (a) Elastic support; (b) bearing fixture; (c) grounded bolt.

2.5.2   Defining the Fixtures to the Chisel

To select the appropriate type of a fixture, you must have an understanding of how the analysed object operates (Figure 2.2). In most cases, the chisel is steadily fixed in the root and loaded at the punching edge. Hence, a **Fixed Geometry** fixture is chosen to be applied to the face in the root of the chisel. The path of commands is described below:

Fixtures (⬚, right click) → Fixed Geometry (⬚) → picking the face in the root of the chisel by direct click on it in the Graphics area → OK (✓, Figure 2.40)

All necessary stages of setting the **Fixed Geometry** fixture are explained in detail later.

Right click the **Fixture** line in **SW Simulation** analysis tree (1, Figure 2.40a).

The **Fixed Geometry** icon is highlighted, that is, the command is active (2, Figure 2.40b). Otherwise, you can simply click on it.

Left click directly in the **Graphics area** at the face where the restraint is applied (3a, Figure 2.40c). The CAD identification of the entity directly appears in the blue window at the left (3b, Figure 2.40c).

Left click on the **OK** icon (✓, 4, Figure 2.40d) to confirm the command and its options.

The **Fixture** property manager closes.

In this section, we were reminded of the DOFs of the analysed objects depending on the type of the object (a solid body, a shell or a beam) and on the entities attached to the applied restraint and the visualisation of different restraints in the Graphics area. We discussed different fixtures depending on the way they are systematized by the software and how they can be introduced to the studied model.

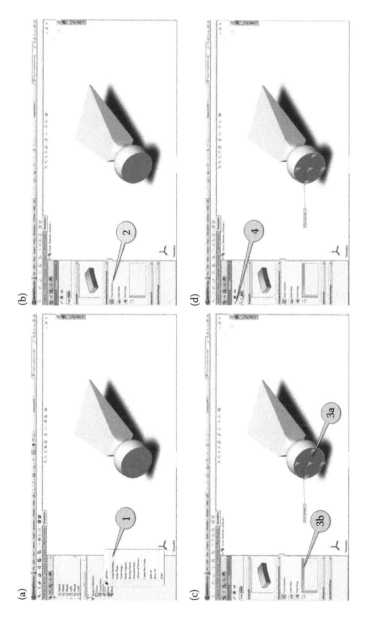

Figure 2.40

Setting a Fixed Geometry fixture at the root of the chisel. (a) Activating the Standard fixture window; (b) choosing a Fixed Geometry command; (c) selecting the face where the restraint is applied; (d) confirming of the fixture commands.

Up to now we learned

- How to introduce a fixture to the model
- How to change symbol settings through which the software describes the fixtures and other restraints
- The main properties of the fixtures, supported by SW Simulation
- How the fixtures are grouped and systematised by the software
- What are the mandatory stages of setting a fixture to the studied object

2.6 INTRODUCING THE LOADS TO THE BODY

2.6.1 Different Structural Loads, Which Can Be Introduced by SW Simulation

The fourth and the last stage, which has to be passed through in order to transform the ready CAD model into a model ready for FE analysis, is applying the loads. As far as the FE method is concerned, the loads have a lot in common with the fixtures. The software enables their automatic adjustment to any change in the geometry of the model. They can be introduced to the model either through the **SW Simulation** command bar or through the **SW Simulation** analysis tree (Figure 2.22) by clicking on the icon **Loads** (). There are two ways to activate the **Load** property manager (Figure 2.41).

In general, the loads can be divided in two major groups – **structural loads** and **thermal/flow loads**. As structural loads are directly related to our study analysis, only them will be discussed in detail here. They are given as follows:

- **Force** (, Figure 2.42a). The forces can be applied to any vertex or point (), joint (), beam (), edge, face or plane (), and they are easily selected in the **Graphics area** by picking directly over the model. The force can be

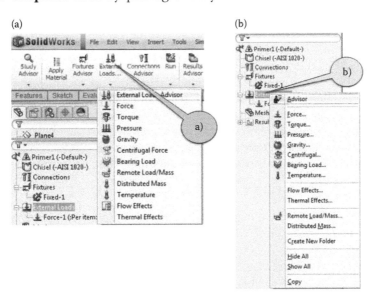

Figure 2.41
Activation of the Load property manager. (a) Activating the Load property manager through SW Simulation command bar; (b) activating the Load property manager through SW Simulation analysis tree.

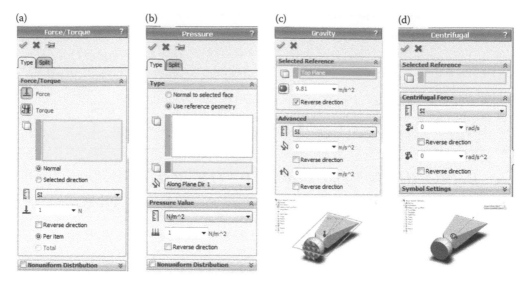

Figure 2.42
Load Property managers – part I. (a) Force/Torque property manager; (b) Pressure property manager; (c) Gravity property manager; (d) Centrifugal property manager.

normal to a face of a solid body or a shell. Then the specified value represents the magnitude of the force. Additionally, **directional forces** can be applied. They act on points, faces, joints, beams, edges and faces. When directional force is applied, it needs a reference mark, which can be an entity (edge or face) or reference geometry (axis for direction or plane). The values of the directional force are introduced by specifying at least one of the following components:

- If the selected entity is a face or a plane – by two tangential (⬚, ⬚) and one normal (⬚) to the entity components
- If the selected entity is an axis – by one radial (⬚), one circumferential (⬚) and one axial (⬚) components
- If the selected entity is an edge – the force acts along the edge (⬚)

When there are some selected entities, the program specifies the value of each definite entity (**per item**) or the **total** value of the force for all selected entities and then distributes the force proportionally among them.

The force can be distributed uniformly or non-uniformly over a face. If the force is non-uniform, a coordinate system can serve as a reference mark. The relative law of the force distribution is a **second-order polynomial** of type

$$F(x, y) = A + B * x + C * y + D * x * y + E * x^2 + F * y^2;$$

where F(x, y) is the relative magnitude of the force applied at a point with coordinates x and y in the selected coordinate system, and A, B, C, D, E and F are polynomial coefficients. When a non-uniform force is applied to a beam, it can be distributed according to a triangular (⬚), a parabolic (⬚), an elliptical (⬚) or a table-driven law. If the load reverses its direction along the geometric entity (face or edge), it is recommended to use the **Split** command (to be explained in detail later).

For dynamic studies, the force can vary in time also.

- **Torque** (⊞, Figure 2.42a). The same **Force/Torque** property manager is used. The torque is applied to faces (usually circular or cylindrical faces) or beams, which can be selected either by directly picking them in the **Graphics area** or by clicking them in the floating **SW design tree** in the **Graphics area**. The reference entities can be either an axis or an edge or a cylindrical face. The value of the torque is directly specified in the **Force/Torque** property manager. The torque can be uniformly distributed only.
- **Pressure** (⊞, Figure 2.42b). The pressure can be applied to any face of a solid body or of a shell or to any edge of a shell by clicking directly on it at the model. The pressure can be either normal to the loaded face or in any other direction. If the direction of the pressure is not normal to the face, a reference entity has to be defined. It can be as follows: a planar face or a reference plane, then the pressure component can be tangential (⧄, ⧅) or normal (⧄) to it; a cylindrical face or a reference axis, then the pressure component is either radial (⧉) or circumferential (⧉) or axial (⧉); an edge (⚊) – the pressure acts along the edge and is introduced either as a positive or a negative value.

  The pressure can be either uniformly or non-uniformly distributed.

  If we apply a **uniform pressure** of a value of p to a face of area A_1, the **equivalent force** will be $P = p * A_1$. If the geometry of the face is modified and the area is set to A_2, then the value of the equivalent force automatically changes to $P = p * A_2$.

  If you prefer to keep that value constant, it is better to apply a force that has a value P. Then even after certain changes in the face geometry, the total value of the force will be preserved.

  If a non-uniform pressure is applied, the law is associated with a previously defined reference coordinate system. It is a second-order polynomial

  $$p(x, y) = V * (A + B * x + C * y + D * x * y + E * x^2 + F * y^2);$$

  where p(x, y) is the magnitude of pressure applied at a point with coordinates x and y in the reference coordinate system; V is the value specified in the **Pressure value** field (⊞); and A, B, C, D, E and F are polynomial coefficients.

  If the pressure reverses or changes its direction, it is recommended to use the **Split** command.

  For dynamic studies, the pressure can vary in time.
- **Gravity** (⊙, Figure. 2.42c). The **Gravity** property manager applies linear accelerations, which distribute over the entire volume of the body. The load value is calculated as the density of the material multiplied by the introduced acceleration. The input values can vary. The directions of the acceleration can be parallel either to the three coordinate axes or to a selected edge. By default, the value of the acceleration is 9.81 m/s², and it is normal to a preselected plane (see the red arrow in Figure 2.42c).

  For dynamic and non-linear analysis, the acceleration can be a time-dependent function.

  The software enables the use of distributed/remote masses (to be discussed later).

- **Centrifugal** (⚙, Figure 2.42d). The Centrifugal property manager applies angular velocity and acceleration to the body. The model spins around the specified axis (Axis 1, Figure 2.42d). The software calculates the centrifugal loads based on the specified values of angular velocity (⚙) or angular acceleration (⚙) and the density of the material. The centrifugal load symbol (⚙) is shown at the centre of gravity of the model (Figure 2.42d).

  For non-linear analysis, the velocity and the acceleration can be time-dependent functions.

Some more structural loads are as follows:

- **Bearing** (⚙, Figure 2.43a). Bearing loads can be applied through the **Bearing** property manager at contacting cylindrical faces or edges of circular shells. The software enables a choice between **sinusoidal** and **parabolic** distribution of the pressure at the interface of contact.
- **Temperature** (⚙, Figure 2.43b). Thermal boundary conditions can be prescribed to faces, edges or vertexes. It is enough to select the entity and to introduce the temperature.

  For nonlinear or transient thermal studies, the temperature can vary with time.
- **Flow effects** (Figure 2.25c). To use this command, an **SW Flow Simulation** should be done in advance, and the loading from the output results file can be imported directly from the static analysis performed by SW Simulation.
- **Thermal effects** (Figure 2.25c). The option enables considering the thermal effects in static studies. To guarantee the success of the analysis, it is mandatory to introduce the coefficient of thermal expansion for each material in the model. This effect is worth studying when there is uniform change in the temperature for the whole model, when there is a results file from previously done thermal analysis or from **SW Flow Simulation** and in some other cases.
- **Remote load/mass** (⚙, Figure 2.43c). Remote loads, restraints and masses are used to simplify the model. There are three basic options to define a remote entity:
  - **Load (direct transfer)**. This option is appropriate when the displacements of the suppressed body are small. The location of the load is specified through the coordinates of the point on the global coordinate system or in a user-defined one. The software calculates the loads at all selected entities within the analysed model.
  - **Load/mass (rigid connection)**. It is used to define forces, moments and masses that are suppressed or not defined in the geometric model. The forces and the moments are applied at remote locations. As these remote entities are outside of the model, their coordinates in the initial or in the user-defined coordinate systems introduce them. The function assumes that all geometric entities connected to the remote location act as rigid ones. If the stress–strain distribution is of no interest, the body can be replaced by a remote mass. Then only its effect on the rest of the structure is analysed. The remote mass is situated at the centre of gravity of the suppressed body.
  - **Displacement (rigid connection)**. This option replaces bodies that can be considered as rigid bodies and their displacement is known. The software calculates the effect of that constraint on the rest of the structure assuming rigid bar connections to all pre-selected entities, that is, faces, edges or vertexes.
- **Distributed mass** (Figure 2.43d). This function is used to simulate the effect of bodies that are suppressed or not included in the modelling when their mass can

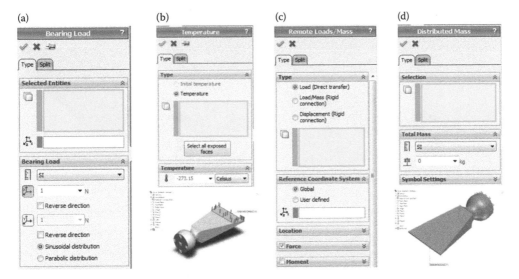

Figure 2.43
Load Property managers – part II. (a) Bearing property manager; (b) Temperature property manager; (c) Remote load/mass property manager; (d) Distributed mass property manager.

be assumed to be uniformly distributed on the specified faces. It is assumed that the suppressed body lies directly on the selected faces, so rotational effects are not considered. To use that command, either gravity or centrifugal effects should be defined. The software distributes the mass proportionally to the area of all selected faces.

Usually there is more than one structural load applied to the analysed model. As there have been assumed static loading and linear stress–strain distribution, the software superimposes (adds) all pressures, forces and remote loads. On the contrary, the software allows the definition of one gravity and one centrifugal load.

2.6.2 Defining the Loads to the Chisel

Based on the shown applications (Figure 2.2) of the chisel, two types of loading are studied:

- **First scenario**: A pressure load distributed over the cutting face of the chisel
- **Second scenario**: Pressure loads distributed over the cutting edge of the chisel and over a section of its side edge

Detailed explanation of both scenarios is provided further:

- **First scenario** (Figure 2.44). In this scenario, the use of the chisel is similar to the applications shown in Figure 2.2c and d. According to experimental data, this load is non-uniform along the longer edge of the face, because of the existing usually omitted friction forces between the cutting and the cut objects. The use of a parabolic function to define the distribution of normal pressure will guarantee high-enough accuracy of our model.

    The external load will be input through the **Force** property manager:

    SW Simulation analysis tree → External Loads → Force (⬇)

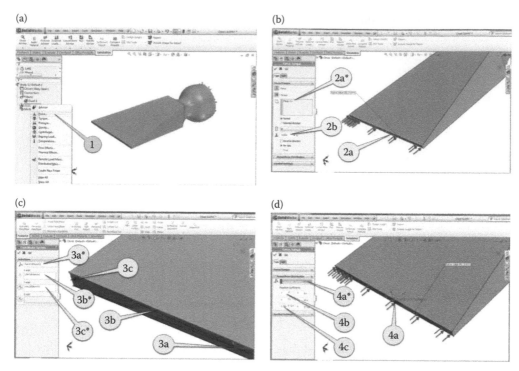

Figure 2.44
Introducing a load normal to the cutting face – first scenario. (a) Opening of the Force property manager; (b) introducing the loaded entity and the direction of the load; (c) definition of a new coordinate system; (d) introducing the load function.

To do so, right click the **External loads** line and pick **Force** from the pop-down menu (picture, 1, Figure 2.44a). It is preferred to use the **Force** load instead of **Pressure**. Thus, the total force value will be constant and will always be equal to the value input in the appropriate window (⊥, Figure 2.44a and b). The force acts at the cutting face. Left click on that face in the **Graphics area**. The face turns blue and its signature automatically appears in the blue window at the left (2a, 2a*, Figure 2.44b). As the load is normal to the face, the radio button below the blue window has to be checked. The force value is 50,000 N (2b, Figure 2.44b). It is the value of the equivalent force of the non-uniformly distributed load.

Further we have to set the coefficients of the parabolic law. We start with the definition of a new coordinate system (Figure 2.44c). The used command path is

Features → Reference geometry → Coordinate system (⌊↵)

At first, we set the origin (↳) of the coordinate system as a mid-point of the diagonals of the rectangle (3a, Figure 2.44c). After that, we set the coordinate axes **X** (3b, Figure 2.44c), which is parallel to the longer edge of the rectangle, and **Y** (3c, Figure 2.44c), which is parallel to the shorter edge of the rectangle, by picking in the **Graphics area** the mid-point and the corresponding lines. Their signatures immediately appear in the appropriate windows of the **Coordinate system** property manager (3a*, 3b* and 3c*, Figure 2.44c). Finally, click **OK**.

Next, we introduce the relative load function. The first step is to select the reference coordinate system (4a, 4a*, Figure 2.44d). According to that coordinate system, the relative load function is 1+0*X+0*Y+0*XY-0.000177*X^2+0*Y^2. This function provides a uniform distribution of the load along the shorter edge and a parabolic distribution along the longer edge. The value of the relative load function is near zero at both side edges and is equal to 1 at the midline. Input the values of the function coefficients in the appropriate fields: 1 (4b, Figure 2.44d) and $(-1/75^2) = -0.0001778$ (4c, Figure 2.44d).

Finally, click **OK** to close the **Force** property manager.

Before continuing to explore different ways of introducing the loads, we rename the current study to **Study_1_Scenario_1** either by right clicking on the **Study** tab at the bottom of the working area and selecting **Rename** from the pop-up menu (Figure 2.45a), or by double left clicking on the **Study** tab. Despite the chosen path, the tab is activated and we can write the new name directly in the **Study** tab (Figure 2.45b). The software automatically updates the new name wherever this is necessary. The new name of the file corresponding to the first scenario is assumed to be **Study_1_Scenario_1**, and this is the name of the **Solution manual** file.

- **Second scenario** (Figure 2.48). Such type of loading can be used to simulate the chisel applications shown in Figure 2.2a and b. Half of the bottom side of the chisel is loaded by forces tangential to the face and in parallel direction to the axis of the chisel. The cutting face is loaded with pressure in the normal direction. Considering the fact that this is a study example, we can assume a uniform pressure distribution over the loaded faces.

To be ready for further analysis and to keep the entire data (geometry, material, fixtures) introduced in the first scenario, we will duplicate them. To do so, we right click the Study tab of the first scenario, titled **Study_1_Scenario_1**, and pick **Duplicate** from the pop-up menu (Figure 2.45a). By default, the new name is **Copy of (the old name of the study)** (Figure 2.46a). But we will change it to **Study_1_Scenario_2** by directly writing the new suggestion in the window (Figure 2.46b). After that, we click **OK** to generate a study with the new name and keep the properties of the duplicated one. To activate the new study, we click on the Study tab named **Study_1_Scenario_2** and the bottom of the working area.

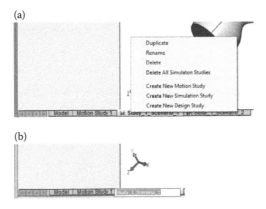

Figure 2.45
Renaming an existing study. (a) Starting the pop-up menu; (b) writing the new name.

(a) (b)

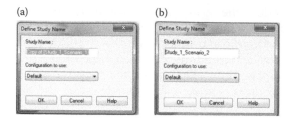

Figure 2.46
Duplicating an existing study. (a) Define Study Name property manager before writing the name of the new study; (b) define Study Name property manager after writing the name of the new study.

Before continuing with the input of the new loads, we delete the applied forces that are transferred to **Study_1_Scenario_2** through (Figure 2.47):

External Loads → Force-1 (right click) → Delete (✕)

Now, we can continue with introduction of the loads from the **second scenario** following the consequence:

- Introducing a uniform pressure over a section of the side edge of the chisel and over its cutting face is done through the path

Fixtures → External Loads → Pressure (⊔⊔⊔)

To input the pressure on the bottom side of the chisel, the following stages must be fulfilled:

- Open the **Pressure** property manager – Right click on the **External loads** and pick **Pressure** in the newly opened window (1, Figure 2.48a).
- Open the **Split** property manager – Click the **Split** tab (2a, Figure 2.48b). Click the **Create Sketch** button to draw a sketch to split the selected face (2b, Figure 2.48b).
- Generate the sketch to be used to split the face – Use the **Sketch** toolbar, particularly the **Line** icon (◥, 3a, Figure 2.48c), to draw a quadrilateral with vertexes at the mid points of the sides and the vertexes of the face (3b, Figure 2.48c). Switch off the **Exit Sketch** button (3c, Figure 2.48c).

Figure 2.47
Deleting an exiting external load.

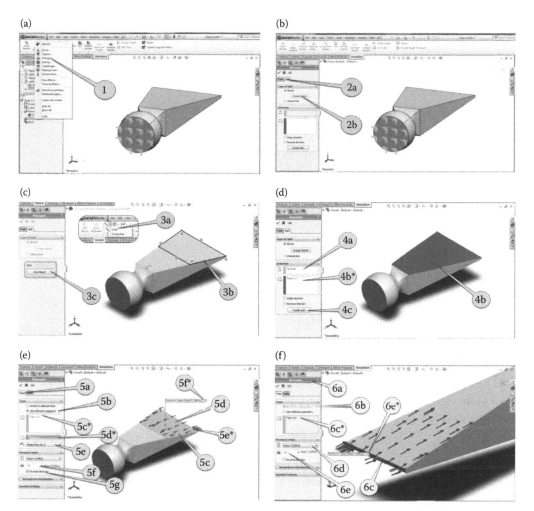

Figure 2.48
*Introducing both pressure loads – second scenario. (a) Starting of the Pressure property manager;
(b) split property manager; (c) generation of the sketch to be split; (d) splitting the face; (e) defining the pressure over the split face; (f) defining the pressure over the cutting face.*

- Split the face – The drawn sketch automatically appears in the **Contour** window (4a, Figure 2.48d); then click the face to be split (4b, Figure 2.48d) and
the software colours it in violet, while its signature automatically appears in
the violet window at the left (4b*, Figure 2.48d). Finally, click the **Create Split**
button (4c, Figure 2.48d).
- Define the pressure over the split entity – Click the **Type** tab (5a, Figure 2.48e).
As the pressure will be tangent to the face, select **Use reference geometry**
(5b, Figure 2.48e). Pick the section of the face to be loaded directly in the
Graphics area (5c, Figure 2.48e); it colours in blue and its signature is automatically written in the blue window on the left (5c*, Figure 2.48e).
- Define a reference entity – Select the same face by clicking on it (5d, Figure
2.48e) and it is directly accepted by the program as a reference entity, whose
signature is in the pink window on the left (5d*, Figure 2.48e). Then introduce

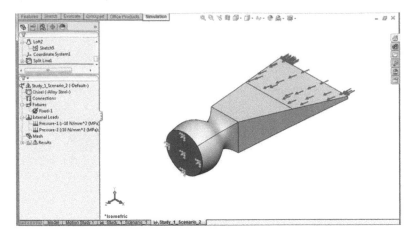

Figure 2.49
Pre-processed model of the chisel – second scenario.

the direction of the pressure by selecting the **Along Plane Dir 2** option (5e, Figure 2.48e), and the direction of the symbols changes as it should be (5e*, Figure 2.48e). Introduce the value of 10 MPa (5f and 5f*, Figure 2.48e). Pick the **Reverse Direction** button to select the correct direction of the pressure (5g, Figure 2.48e). Click **OK** to confirm the input pressure.

- Define the pressure over the cutting face – Open the **Pressure** property manager (1, Figure 2.48a and 6a, Figure 2.48f). Select the option **Normal to selected face** (6b, Figure 2.48f). Click the loaded face in the **Graphics area** (6c and 6c*, Figure 2.48f). Choose the units – MPa (6d, Figure 2.48f) – and introduce the value of the pressure, which is assumed to be equal to 10 (6e and 6e*, Figure 2.48f). There is no need to select the **Reverse Direction** button. By default, the **Normal to selected face** loads point at the face.
- Click **OK** to close the **Pressure** property manager.

Starting the processing model of the chisel of scenario 2 is shown in Figure 2.49.

We studied different types of loads and how to start the Loads property managers. We commented on how the loads are applied to the model and what is the difference between Force and Pressure loads.

We have learnt how to

- Start Loads property manager
- Introduce force and pressure loads
- Introduce loads on the entire entity or how to split the entity if necessary
- Define uniform and non-uniform loads

CHAPTER 3

DEVELOPMENT OF A FINITE ELEMENT MODEL OF A BODY (PROCESSOR STAGE)

3.1 HOW DOES FINITE ELEMENT ANALYSIS WORK?

The core of the **finite element analysis (FEA)** is to divide the solid body model (Figure 3.1a) into a lot of small pieces, with simple shapes, called **finite elements (FEs)** (Figure 3.1b). They are connected at common points called **nodes**. The **FE method (FEM)** predicts the behaviour of the model, based on the equations describing the behaviour of each FE as well as the inter-relations among them and their interaction with the ambient environment.

The very process of dividing the model into FEs is called **meshing**, and the result of that division is a mesh of FEs. Meshing is crucial for the final success of the FEA. While meshing a solid body model, SW Simulation generates two types of FEs (Figure 3.2):

- **Linear solid element** (Figure 3.2a) is a linear tetrahedral FE. It is defined by four nodes at its corners that are connected through six straight edges. As the displacement functions along these edges are linear functions, the element is also called an **FE of the first order**. Using this type of FE or FEs, the program generates a **draft quality mesh**.
- **Parabolic solid element** (Figure 3.2b) is a parabolic tetrahedral FE. It is defined by 10 nodes, situated at the corners as well as at the mid-points of the edges. The connecting edges are parabolic curves. The displacement functions along the FE edges are parabolic functions as well; thus, these FEs are also known as **second-order FEs**. Using them, the program generates a **high-quality mesh**.

The use of linear or parabolic solid elements, as well as their size, strongly influences the accuracy of the analysis. This impact will be discussed in detail later.

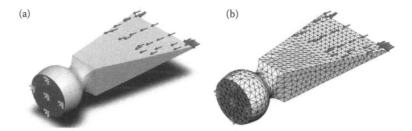

Figure 3.1
Meshing a solid body. (a) Solid body model; (b) FE model.

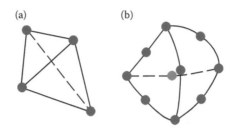

Figure 3.2
Solid FEs (SW Simulation on-line help). (a) Linear solid element; (b) parabolic solid element.

The software was also operated with 2D and 1D FEs. Their applications will be studied further when shells and structures of beams are analysed.

We explained the main idea of the FEM, that is, the object to be divided in small pieces called FEs, which are connected to each other by nodes. The very process of dividing the object into FEs is called meshing, and the result is a mesh of FEs.

In this section, we learned

- What are finite elements and what are nodes
- What meshing is
- What are the two types of 3D FEs, which are supported by the software and what are the main differences between them

3.2   WHAT ARE THE FEs AND THE MESH?

Meshing can be started either through the command bar or through the analysis tree (Figure 3.3). To start meshing through the command bar, you must right click the **Run** icon (⬚) and choose **Create Mesh** (⬚) in the pop-down menu (Figure 3.3a). If you decide to use the **Mesh** command (⬚) at the **SW Simulation** analysis tree, you will have access to a more detailed pop-up menu (Figure 3.3b).

Starting the meshing through the **SW Simulation** analysis tree provides access to more options with this procedure. We can see that the commands through which we can control the meshing process are divided into a few groups.

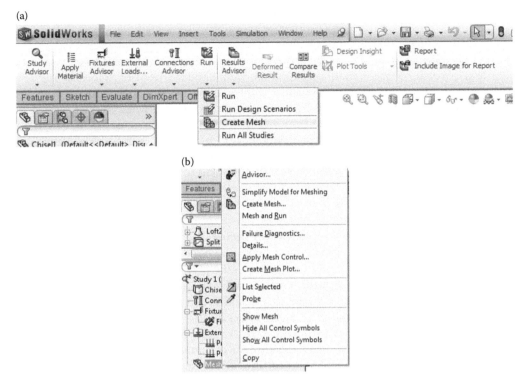

Figure 3.3
Starting the mesh generation. (a) From the command bar; (b) from the analysis tree.

The first group unites the commands that help the generation of the mesh. They are the following:

- **Simplify model for meshing** (⚙, Figure 3.3b). Another way to activate this utility is (Figure 3.4)

$$\text{Tools} \rightarrow \text{Find/Modify} \rightarrow \text{Simplify...}$$

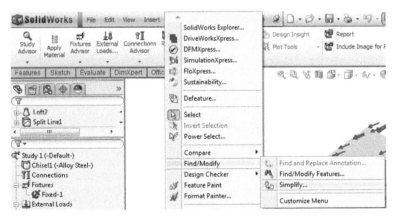

Figure 3.4
Starting the Simplify command through the SW command bar.

This utility is helpful when there is an internal calculation of "insignificant volume" based on the size of a part or assembly. The selected features below that volume are suppressed.

The **Simplify** utility can be applied to fillets, chamfers or holes to extrude and revolve to more than one of them simultaneously. After selecting the impacted **Features** (Figure 3.5a and 1, Figure 3.5b), the **Simplification factor** should be set (2, Figure 3.5b). This will change the insignificant volume.

The **Simplification factor** is used differently according to the selected method (Figure 3.5c): the **Feature Parameter** (3a, Figure 3.5c) or the **Volume Based** (3b, Figure 3.5c), yet it always affects the actual volume of the model or of the preselected feature. The **Feature Parameter** identifies the feature for simplification regarding $P < \sqrt[q]{(V_{min} * S)}$, where P is the value of the main feature parameter (e.g. for a hole this is the diameter and for a fillet this is the fillet radius); V_{min} is the minimum volume of the body associated with the feature, and S is the value of the **Simplification factor**. The **Volume Based** method identifies the feature for simplification under the condition $V < CV * S$, where V is the feature volume and CV is the volume of the part or of the assembly. For an assembly file, the option **Ignore features affecting assembly mates** can be picked so those features that would cause mate failures are not suppressed.

After clicking **Find Now** (4, Figure 3.5d), a tree of features with insignificant volumes is displayed in the **Results** section (5, Figure 3.5e). When **Create derived configurations** is unsuppressed, the simplified features can be added to a different configuration selected under **Configurations** (6, Figure. 3.5f). Further, this configuration can be renamed in the **Name** box and is updated automatically.

- **Create Mesh** (): This command can be activated either from the **SW Simulation** toolbar (Figure 3.3a) or from the **SW Simulation** analysis tree (Figure 3.3b). Then the **Mesh** property manager opens (Figures 3.6 and 3.8),

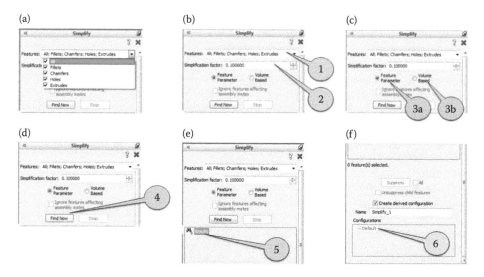

Figure 3.5

Simplifying the model. (a) Features supported by Simplify utility; (b) introducing the value of the Simplification factor; (c) selecting the method of simplification; (d) starting the process of simplification; (e) Results section, where all simplified features are displayed; (f) creation of a new configuration.

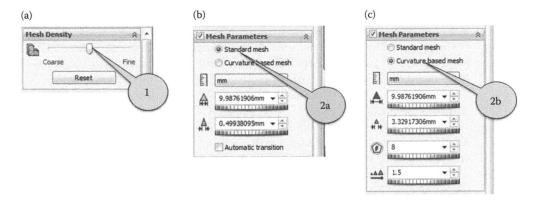

Figure 3.6
Mesh property manager – Mesh Density and Mesh Parameters sub-windows. (a) Mesh Density option; (b) standard mesh parameters; (c) curvature-based mesh parameters.

and all options of the mesh should be input. Of course, there are some options when values are set by default.

- The first to be introduced are the parameters regarding the **Mesh Density** (Figure 3.6a). The slider (1, Figure 3.6a) enables easy simultaneous variations of the size and tolerance of the FEs. By default, it is situated somewhere in the middle of the line. If we move the slider to the left (**Coarse side**), the size of the FEs increases twofold compared to the initial data. Moving the slider to the right (**Fine side**) decreases the size of the FEs and increases the mesh density. As a result, the mesh becomes finer when moving the slider to the right. The global element size can be decreased up to half of its default value.

- The next step is to introduce the **Mesh Parameters**. The program offers two types of mesh – a **Standard mesh** (2a, Figure 3.6b) or a **Curvature-based mesh** (2b, Figure 3.6c). The **Standard mesh** activates the Voronoi–Delaunay meshing scheme, whereas the **Curvature-based mesh** automatically enables the use of FEs of higher order in higher-curvature areas. After that, the **units** of the mesh size and tolerance have to be set (▣, Figure 3.6b and c).

- Both the **Global size** option (▲, Figure 3.6b) and the **Tolerance** option (▲, Figure 3.6b) are available only for a **Standard mesh**. Through the **Global size**, the program suggests a default value (corresponding to the default position of the slider in the **Mesh Density** sub-window) of the FEs, based on the model volume and the surface area. By default, the tolerance value is 5% of the global element size. If the distance between two neighbour nodes is smaller than that, the nodes are merged unless otherwise specified in the contact conditions. The upper limit of the tolerance is 30% of the global element size. Sometimes, adjusting the tolerance helps in solving problems related to the mesh creation. When **Automatic transition** is checked (Figure 3.6b), the program automatically applies mesh control to all fine details of the model. Thus, the generated mesh is finer at the area with small features, and the number of FEs is higher compared to when this option is unchecked.

- All further discussed options are available only for **Curvature-based mesh**. They are **Maximum element size** (⊿, Figure 3.6c), which is used for boundaries with the lowest curvature; **Minimum element size** (⊿⊹, Figure 3.6c), which is used for boundaries with the highest curvature; and **Min number of elements in a circle** (◉, Figure 3.6c), which specifies the number of elements in a circle (Figure 3.7). Thus, if the value of **Min number elements** is set to 8, eight triangles can form an octagon with a side of h, which is to be inscribed in an imaginary circle with radius r. The length of the side of the octagon is $h \approx r\alpha$, where α is the corresponding central angle; in this case $\alpha = 2\pi/8$. This option is effective if the size h is in between the values of the **Maximum element size** and the **Minimum element size**. The **Element size growth ratio** (≜, Figure 3.6c) specifies the global element size growth ratio starting from regions of high curvatures in all directions.
- The next sub-window in the **Mesh** property manager includes the **Advanced** options (Figure 3.8a). **Jacobian points** (Figure 3.8a) is an option that is available for high-quality mesh only, that is, with FEs of second order. The mid-side nodes of these FEs wrap around the geometry of high curvature regions much better than the linear FEs. This option sets the number of integration points to be used in checking the distortion level of tetrahedral elements. The **Jacobian check** can be based on 4, 16, and 29 **Gaussian points** or **At Nodes**. Basically, there is no real hard evidence that using any more than the default amount of points for the Jacobian check makes the mesh much better; however, the higher values increase the resolution of the computation of the Jacobian value, without any substantial increase in time of mesh generation.

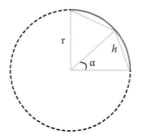

Figure 3.7
How to determine the element size (SW Simulation on-line help).

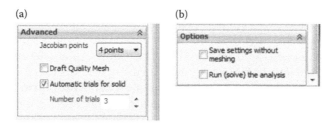

Figure 3.8
Mesh property manager – Advanced and Options sub-windows. (a) Advanced options; (b) Options.

- If the **Draft Quality Mesh** (Figure 3.8a) is checked, 4 corner nodes for each 3D FE and 3 corner nodes for each 2D FE will be used. This option is recommended for models without any curvature surfaces or for quicker calculations.
- The **Automatic Trials for Solid** (Figure 3.8a) is an option available for standard mesh only. If it is checked, the meshing algorithm remeshes the model automatically in case of failure of the previous trial, using smaller global elements with a reduced ratio and tolerance equal to 0.8 for each trial. The option **Number of trials** (Figure 3.8a) sets the maximum number of mesh trials.
- For assembled solid body models, the option **Remesh failed parts with incompatible** mesh is also available. It specifies that the incompatible meshing should be used for bonded bodies if the compatible meshing has failed (to be explained in more detail later).
- The last sub-window in the **Mesh** property manager is **Options** (Figure 3.8b). There are two accessible options that can be kept as they are selected or deselected: **Save settings without meshing** and **Run (solve) analysis**.
- The quality parameters for all FEs can be plotted through the **Create Mesh Plot** command (Figure 3.3b), which has the options **Mesh**, **Aspect ratio** and **Jacobian** (Figure 3.9). The **Aspect ratio** is the ratio of the longest normal to the shortest normal in an FE, where the normal is dropped from the vertex node to the opposite face of the element, that is, the **Aspect ratio** measures how 'stretched out' the element is. It is recommended that the aspect ratio be under about 5 for structural analysis.
- **Mesh and Run** (Figure 3.3b) meshes the model by using the input mesh settings and automatically runs an analysis after that.

The second group of commands in the **Mesh** pop-down menu (Figure 3.3b) unites commands that help in checking the quality of the generated mesh. They have to be activated after the mesh generation; otherwise, they provide no information.

- **Failure diagnostics** (Figure 3.3b) enlists (▤) and highlights all the components that failed to be meshed. These can be components (▨), faces (▨) or edges (▨). **Failure diagnostics** also can be done through **Simulation advisor**. Then the results appears at the **Simulation advisor** on the right side of the working area (1, Figure 3.10a).
- **Details** (Figure 3.3b) automatically opens a window with the entire data related to the just generated mesh (Figure 3.10b). You can see the name of the study, the type and properties of the FEs, the number of nodes and elements, the **Aspect ratio**, the **Jacobian** and the time to complete the mesh. Thus, you

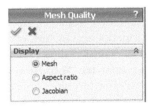

Figure 3.9
Creating of mesh quality plots.

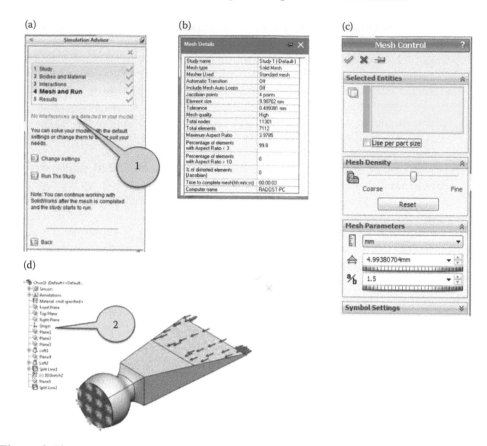

Figure 3.10
Checking the quality of the generated mesh. (a) Failure diagnostics; (b) Mesh Details; (c) Mesh Control property manager; (d) Floating Feature Manager design tree.

can judge the quality of the mesh and decide whether to leave it as it is or to re-mesh the model.

- **Apply Mesh Control** (⬛, Figure 3.3b). It is known that the smaller the size of the FE, the higher the accuracy of the calculation. Yet, this results in more time for the program to find the solution. Thus, it is recommended to look for a balance between the size of the FEs and the precision needed. This command enables the user to use FEs of different size within one and the same model. The **Mesh Control** property manager controls the size of the FEs, denoted by e, and the growth ratio, denoted by r. It can be applied to a selected geometric entity, including a vertex, point, edge, face or any other component. The size of the elements increases, radiating the selected entities, according to the law e, $e * r, e * r^2, e * r^3, ..., e * r^n$. The mesh radiates from vertices to edges, from edges to faces, from faces to components and from a component to connected components. The entities can be selected by directly clicking on them at the graphics area or from the floating **Feature Manager design tree** (Figure 3.10d). After the selection, they are coloured in blue, and their signatures automatically appear in the blue window **Selected Entities** (▢, Figure 3.10c). If the **Use per part size** is checked, the mesh size is calculated based on the individual part size. After selecting the entities, the **Mesh Density** has to be introduced using

56

the slider in the corresponding window. By default, the program calculates the optimum size of the controlled elements based on their volume and sets the slider in the middle of the range. By moving it to the left, the element size is increased twofold, while moving it to the right, decreases the size up to a half. Additionally, the properties of the mesh can be adjusted through the **Mesh Parameters** window. **Units** (▤), **Element size** (⬦) and **Ratio** (%), which sets the ratio between the element size in two neighbouring layers, can be input.

- **Create Mesh Plot** (Figure 3.3b) has been discussed in detail in the previous item (Figure 3.9).
- **List Selected** (▨) and **Probe** (✎) commands open one and the same **Probe Result** window (Figure 3.11), where the program displays the node or the element numbers, and the global coordinates of the nodes or of the element centres.
  - **At location** (Figure 3.11a) displays the results for individual nodes or elements that are picked in the **Graphics area**.
  - **On selected entity** (Figure 3.11b) displays the results for all nodes or elements on the selected entities (faces, edges or vertices).
  - **Distance** (Figure 3.11c) measures the distance between every two nodes, selected in the **Graphics area**.
- The next three commands, **Hide Mesh, Hide All Control Symbols** and **Show All Control Symbols**, adjust the visualisation of the meshed model in the **Graphics area** and do not influence the properties of the mesh.

Finally you must remember that the right meshing is crucial for the accuracy of the analysis. By default, the mesh consists of one type of FEs, unless otherwise specified. The generated mesh based on global element size and local mesh control provides the optimal ratio precision to computer time.

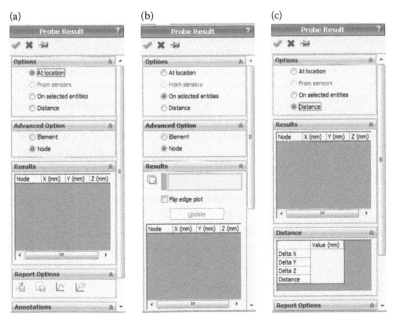

Figure 3.11
Probe Result. (a) At location; (b) on selected entities; (c) distance.

We studied the different mesh types applied to solid body models and how to set the main properties of the mesh, including the maximal and the minimal element sizes. We commented on the main differences between the standard and the curvature-based mesh, as well as how to check the quality of the created mesh. We briefly mentioned what is provided by the program option of Mesh Control, enabling the use of denser meshes in vulnerable areas.

In this section, we learned

- Both types of mesh provided by the program: standard and curvature-based mesh
- The mesh parameters and their impact on the created mesh
- How to control the mesh density varying the maximal element size
- The Mesh control as an option to better the mesh quality
- Some ways to control the mesh quality
- List selection and probe commands

3.3   MESHING OF THE ANALYSED BODY

It has been decided to analyse the chisel loaded by two pressure loads, that is, **Scenario 2** from Chapter 2.

We have already introduced the material, the fixtures and the loads to the geometrical model of the chisel. The next step is to create the mesh, that is, to complete the transformation of the model of the solid body into an FE model. To start meshing procedure:

Mesh (right click) → Create Mesh…(🖺)

Further,

- We assume the **Mesh density** to be the default, that is, the slider to be situated in the middle of the range (1, Figure 3.12a).
- **Standard mesh** is preferred as there are no entities of high curvature in the model. Hence, we set the options of the **Mesh parameters** sub-window (Figure 3.12b) as follows: check the **Standard mesh** (2a, Figure 3.12b); set **Size of the elements** to 10 mm (2b, Figure 3.12b) and the **ratio** to 0.5 (2c, Figure 3.12b). Finally, we check the **Automatic transition** to activate the mesh control.
- The advanced properties of the mesh are defined through the **Advanced** sub-window of the **Mesh** property manager (Figure 3.12c). They are as follows: **Jacobian** is equal to the highest possible value for that program, that is, 29 (3a, Figure 3.12c); **Automatic trials for solid** is checked (3b, Figure 3.12c) and the **Number of trials** is limited to 3 (3c, Figure 3.12c).
- Check **Run (solve) the analysis** in the **Options** sub-window (4, Figure 3.12d).
- Click **OK** to save the mesh settings, to start the meshing of the object and to run the analysis (5, Figure 3.12d).

Before going further, we can verify the quality of the mesh. This step is not mandatory but is recommended:

Mesh (right click) → Details

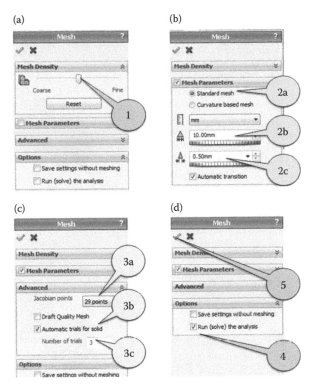

Figure 3.12
Create mesh. (a) Adjusting mesh density; (b) setting mesh parameters; (c) input of the advanced values; (d) finalising the mesh generation and running the analysis.

A window with the details of the FE model appears (Figure 3.13a). This is a mesh of linear solid FEs. The number of elements is 7298, and the number of nodes is 11579 (approximately). For your model, it can vary slightly depending on the software mathematical algorithms. Even more, if we re-mesh the model, some of these data could be changed.

The next command from the pop-down **Mesh** menu is

Mesh (right click) → Create Mesh Plot

which starts the **Mesh quality** property manager, which itself provides the following options:

- *Mesh* – generates the plot of the mesh (Figure 3.13b).
- *Aspect ratio* – shows the aspect ratio values over the entire model (Figure 3.13c). Its maximum is 3.74 at the cutting edge of the chisel.
- *Jacobian* – shows the Jacobian values over the entire model (Figure 3.13d). All FEs are well generated, and there are no Jacobian values higher than 2.18.

Consequently, the solid mesh is well generated and will provide accurate results.

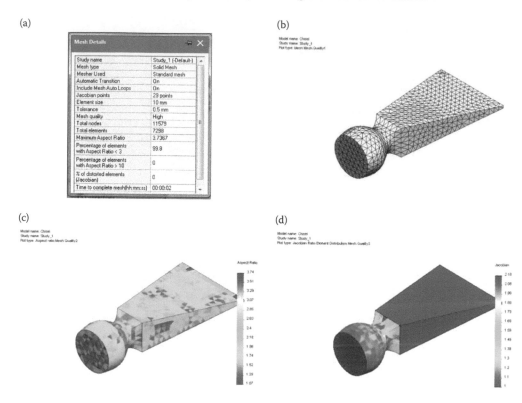

Figure 3.13
Verification of mesh quality. (a) Mesh details; (b) plot the mesh; (c) aspect ratio distribution; (d) Jacobian values distribution.

We meshed the studied chisel, applying a standard mesh with an FE maximal size of 10 mm and a tolerance of 0.5 mm. We provided all data to assess the quality of the created mesh and concluded that, based on the percentage of the Aspect ratio values smaller than 3 and the percentage of distorted elements, the mesh is well generated and does not need further modification.

In this section, we practiced our knowledge of creating a mesh and meshed the chisel using linear solid FEs. We learned

- How to obtain current data about the number of FEs and nodes
- That the percentages of the Aspect ratio and distorted FEs can be used as a criterion of mesh quality
- The necessary computer time to generate the mesh

3.4 RUNNING THE FEA

Finally, it is time to say something about the actual mathematical solution of the problem. We have already discussed the solvers built into the software.

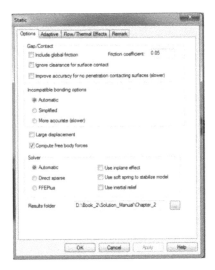

Figure 3.14
Adjusting the properties (the options) of the analysis solver.

The properties (options, features, characteristics) of the analysis solver are set at the pre-processor stage, when the study properties are defined. Yet, if necessary, it can be changed at this stage. For our analysis, the Automatic option for selecting the solver is checked (Figure 3.14):

Name of the analysis in the SW Simulation analysis tree (right click) → Properties → Static → Options → Solver → Automatic → OK

The most important thing at that stage of the analysis is the passive involvement of the user. This significantly reduces the possibility of numerical mistakes; however, it could multiply the impact of the mistakes made during the development of the CAD model or during the pre-processor stage, as well as during the mesh creation.

Additionally, we will discuss briefly the calculation workflow for performing a static linear analysis:

- Based on the input data (geometry, material properties, boundary conditions, etc.), the program performs all necessary mathematical equations. As a result, it calculates the displacement components (translational and rotational) of each node.
- Based on the obtained displacement values, the program calculates the strain components as a ratio of the change to the initial value of the length.
- Finally, based on the calculated strain results and Hook's law, the program calculates the stresses. During the first run, the stresses are calculated at special points, called **Gaussian points**, located inside the elements, in a way that provides optimal numerical results. After that, the results are extrapolated to the nodes of each element. Thus, some different values can appear in a node that is common for a few elements. This is a consequence of the basics of the FEM as a method for approximate numerical calculations. The program presents the stress results in two totally different ways: **element mode** – when the program averages the stress values of all nodes that belong to the element; and **node mode** – when the program averages the stress values of nodes from all elements that share a common node. While the second mode provides a more fluent stress diagram, the first

mode is recommended for checking whether the density of the mesh is appropriate. If the stress diagram in the element mode is fluent, then the mesh density is the right one; otherwise, a finer mesh could improve the resolution.

To run the analysis, we can use any of these three command paths:

- SW Simulation command bar → Run (, Figure 3.15a)
- SW Simulation analysis tree → Mesh (or , right click) → Mesh and Run (Figure 3.15b)
- SW Simulation analysis tree → Name of the study (right click) → Run (, Figure 3.15c)

While the calculation is running, the **Study manager** appears (Figure 3.16b). This provides information about the memory usage, elapsed time and reached percentage of the calculation process (18.2%). Additionally, the numbers of degrees of freedom (33,978), of nodes (11,579) and of elements (7298) are shown. The number of degrees of freedom is equal to the number of the searched displacement components, that is, to the number of equations in the solved linear system.

In Figure 3.15, the **Current task** (Stress Reaction Calculation) and its level of completion (99%) are given. The **Convergence plot** is also provided (Figure 3.16a).

Solver parameters are shown in Figure 3.16c. They can be changed during the calculation. For example, to improve the accuracy, **Input stopping threshold** can be decreased. If there are calculation problems and the converging process is too slow, either the value of the **Input stopping threshold** can be increased or the **Input Maximum number of iterations for the iterative solver** can be decreased, or both.

The next stage is systematisation, visualisation and analysis of the results, that is, the post-processor stage.

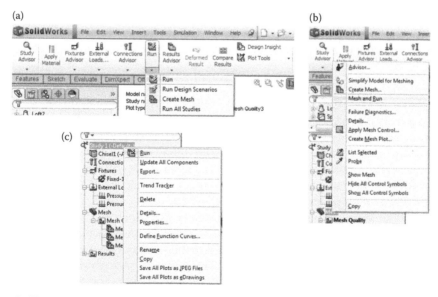

Figure 3.15
Running the analysis. (a) From the command bar; (b) from the analysis tree – I version; (c) from the analysis tree – II version.

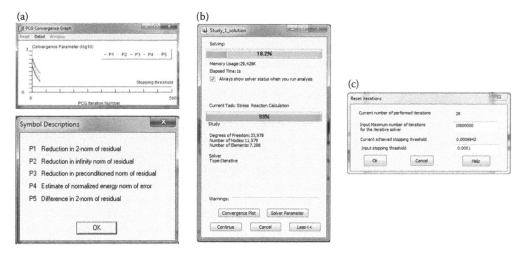

Figure 3.16
Analysis windows. (a) Convergence plot; (b) study window; (c) solver parameters.

We studied the workflow of the run analysis. We discussed how to assess the feedback provided by the Study manager and how to re-define during the calculations the convergence criteria, if necessary.

In this section, we learned

- How to run the analysis
- How to re-define the used solver
- How to modify the convergence options if necessary
- How to make the calculations run quicker, reducing the accuracy of the analysis

CHAPTER 4

VISUALISING AND SYSTEMATISING THE RESULTS OF FEA (POST-PROCESSOR STAGE)

4.1   SETTING THE ANALYSIS AND THE RESULTS PREFERENCES

Anytime before starting the analysis, we can set the preferences that "tell" the program how to visualise the obtained results. Otherwise, it uses the default settings. All analysis preferences, including the visualisation of the results, are set in the **Options** property manager. It is activated through the path

Simulation → Options (Figure 4.1a)

The **General System Options** include information about the type of the displayed massages; the quality of the load and fixture symbols – wireframe or shaded; the colours of the mesh; whether to start or not the **Simulation Advisor**; whether to hide or not the excluded components from the analysis; and the font properties (Figure. 4.1b).

As far as the visualisation of the results is concerned, the **Result plots** buttons (Figure 4.1c) can either be selected or not. The choice affects all result plots. There are five options to consider:

- **Dynamic plot update** – it enables automatic updates of the plots as the parameters that control the plots' appearance are modified. It is recommended to switch this option off to improve performance of viewing results for large models.
- **List result quantities in higher precision** – it enables the results to be listed with up to 16-digit precision.

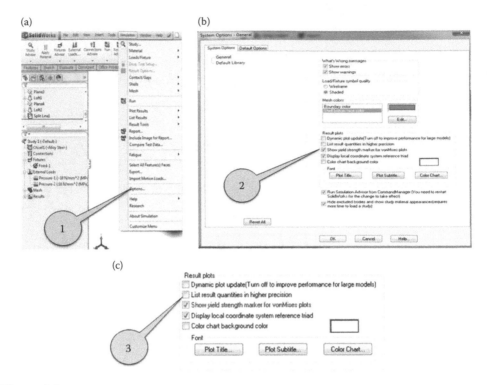

Figure 4.1
SW analysis options. (a) Starting Options property manager through the command bar; (b) System Options property manager; (c) Result plots sub-window.

- **Show yield strength marker for von Mises plots** – when selected, the location of the material yield strength on the plot legend is marked with a red arrow.
- **Display local coordinate system reference triad** – it displays the local coordinate system reference triad at the lower right corner of the graphics area.
- **Color chart background color** – sets the background colour for the plot legend.

There are some options in the **Default Options** window that influence the entire analysis.

- **Units** (Figure 4.2a) – the user can choose among the following systems: **SI**, **English** or **metric**. The SI unit system will be used for all exercises in this book (1a, Figure 4.2a). After that, the basic units for **Length/Displacement**, **Temperature**, **Angular velocity** and **Pressure/Stress** are defined (1b, Figure 4.2a).
- **Load/Fixture** (Figure 4.2b) – the user defines the size (2a, Figure 4.2b) and the colour (2b, Figure 4.2b) of the used symbols and can choose to select or not the preview option (2c, Figure 4.2b).
- **Mesh** (Figure 4.2c) – defines the quality of the mesh – **Draft** or **High quality** as well as the **Jacobian** points (3a, Figure 4.2c). Further, the mesh settings are set (3b, Figure 4.2c). The influence of these settings on the accuracy of the

(a)   (b)

(c)   (d)

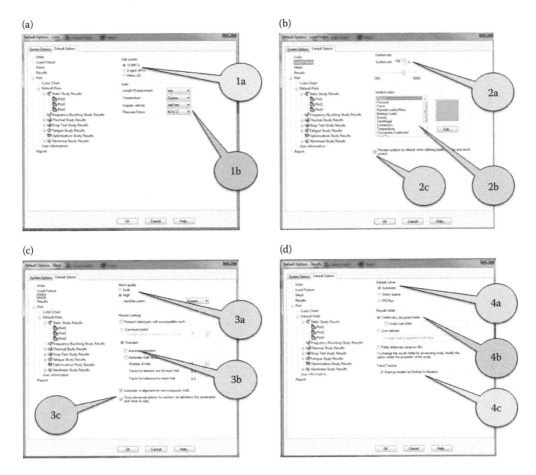

Figure 4.2
Default Options. (a) Units Default Options; (b) Load/Fixture Default Options; (c) Mesh Default Options; (d) Results Default Options.

results has been discussed in the previous item. Finally, both last options can either be activated or not (3c, Figure 4.2c).

- **Results** (Figure 4.2c) – here the default solver should be chosen (4a, Figure 4.2d). The directory where the results will be saved is directed (4b, Figure 4.2d). It is recommended to select the last button to enable the backing-up of the model for future restoration (4c, Figure 4.2d).

From here on, the **Plot** options (Figure 4.3) will be discussed in detail. These settings are directly related to the visualisation of the obtained results. Of course, the plots are not limited to the default ones, but it is better if the initial plots present the most common results.

- **Plot** (Figure 4.3a) – through that window, **Annotation and range** properties are set (1a, Figure 4.3a). By either checking or not the buttons, the user chooses whether to show the minimal and the maximal values as well as the range of the displayed components. Additionally, some settings' options regarding the fringe (Point, Line, Discrete and Continues) and the boundary (None, Model, Mesh

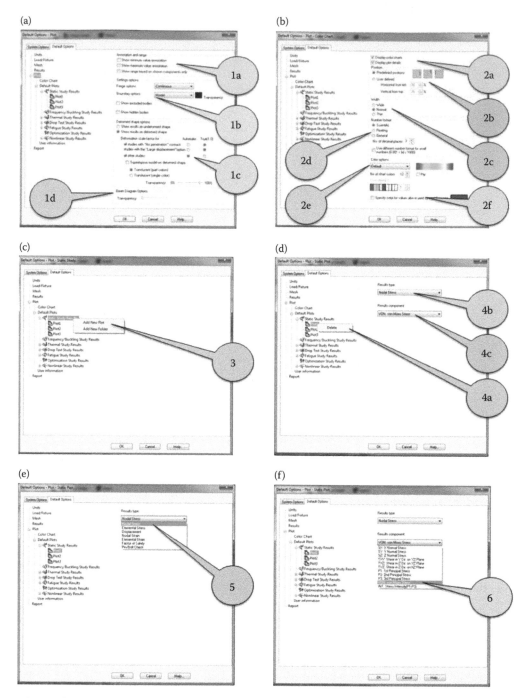

Figure 4.3
Options of the plots. (a) Plot options; (b) Color Chart options; (c) adding new plots or folders to the results tree; (d) Results tree; (e) Results type; (f) Results component.

and Translucent) presentations and displaying the hidden and excluded components from the analysis can be defined (1b, Figure 4.3a). The next sub-panel enables the introduction of the deformed shape options (1c, Figure 4.3a), while the last sub-panel is related to the display of the beam diagrams if they exist (1d, Figure 4.3a). For example, for a solid body analysis, as an example provided here, there are no beam parts; hence, there are no such diagrams.

- **Color Chart** (Figure 4.3b) – Through that panel, we can choose to either display or not colour charts and plot details (2a, Figure 4.3b), and to define the position and the width of the chart (2b and 2c, Figure 4.3b). The format of the plot numbers can be Scientific (1.234e+01), Floating (1.234), General, where the number of decimal places is set (if this number is 2, the number is 1.23), or any appropriate, according to the programme for small numbers (2d, Figure 4.3b). The next is the choice of colour palette (2e, Figure 4.3b). It can be **Default** (▬▬▬), **Rainbow** (▬▬▬), **Grayscale** (▬▬▬) or **User Defined**. The colour of all areas where **von Mises** stresses are above the yield stresses can also be specified (2f, Figure 4.3b).

- If you right click on a certain type of analysis, **Static Study Results** for example (3a, Figure 4.3c), you can **Add New Plot** to the group of the pre-defined plots related to the folder or **Add New Folder** for a new group of plots. Thus, the **Results** tree can be managed.

- If you right click on a certain plot, you can **Delete** it from the results tree (4a, Figure 4.3d), or if you left click on a certain plot, you can simply change its properties, regarding **Results** type (4b, Figure 4.3d) and **Results** component (4c, Figure 4.3d).

- All possible **Results types** can be seen in Figure 4.3e. The corresponding **Results components** are given in Figure 4.3f. The program automatically relates the components to the chosen type and thus prevents wrong choices. This relation is given in Table 4.1.

- The **User Information** panel (Figure 4.4a) enables the input of some data related to the names of the company and the user and the import of the company logo as an image file of the type *.bmp or *.jpg. This information will be written in the generated report if the last option is selected.

- The final report generation is of high importance for the final presentation of the project results. The program helps us with this uneasy task. The basic sections that should be included in the report can be selected from the **Report** panel (Figure 4.4b). It is crucial at this point to select the right type of analysis from the **Report formats** menu, because the available report formats are customised according to the type of study. Report sections depend on the chosen type of the analysis, and some of them, which are typical for **Static analysis report**, can be seen in the figure (Figure 4.4b). Some user-defined comments and additional data not included in the format (e.g. addition of images of all plots to the **Study Results** section) can be added to each **Report section**. This is free text information, written in the window on the right side of the **Report** panel. All information provided by default for each item of the **Results section** for **Static Study** report is systematised in Table 4.2. Sections included in the final report are not limited to those in the table. An additional user-defined section can be embedded in the **Appendix section**. It can be imported as MS Office data, as a SW image or as any other image file.

Table 4.1

Relation between Results Type and Results Components

Results Type		Results Component
Nodal Stress Elemental Stress	Nodal stress component Elemental stress component	• **SX**: X normal stress • **SY**: Y normal stress • **SZ**: Z normal stress • **TXY**: shear in Y direction on YZ plane • **TXZ**: shear in Z direction on YZ plane • **TYZ**: shear in Z direction on XZ plane • **P1**: 1st principal stress • **P2**: 2nd principal stress • **P3**: 3rd principal stress • **VON**: von Mises stress • **INT**: stress intensity (P1–P3) • **ERR**: energy norm error • **CP**: contact pressure
Displacement	Displacement component	• **UX**: displacement (X direction) • **UY**: displacement (Y direction) • **UZ**: displacement (Z direction) • **URES**: resultant displacement • **RFX**: reaction force (X direction) • **RFY**: reaction force (Y direction) • **RFZ**: reaction force (Z direction) • **RFRES**: resultant reaction force • **RX**: rotation (X direction) • **RY**: rotation (Y direction) • **RZ**: rotation (Z direction) • **RMX**: reaction moment (X direction) • **RMY**: reaction moment (Y direction) • **RMZ**: reaction moment (Z direction) • **RMRES**: resultant reaction moment
Nodal Strain Elemental Strain	Nodal strain component Elemental strain component • **SEDENS**: Strain energy density • **ENERGY**: Total strain energy	• **EPSX**: X normal strain • **EPSY**: Y normal strain • **EPSZ**: Z normal strain • **GMXY**: shear in Y direction on YZ plane • **GMXZ**: shear in Z direction on YZ plane • **GMYZ**: shear in Z direction on XZ plane • **ESTRN**: equivalent strain • **E1**: normal strain (1st principal direction) • **E2**: normal strain (2nd principal direction) • **E3**: normal strain (3rd principal direction)
Factor of Safety Pin/Bolt Check	No result components	

The next step is the input of the **Header** information in the **Report** window. It includes Designer, Company, URL and Logo (if any), Address, Phone and Fax. All these data are optional and can be omitted in the final report. With regards to saving the generated file somewhere, the **Report folder** should be set. By default, it is the **Results folder**, although the user can change the directory path.

(a)

(b)

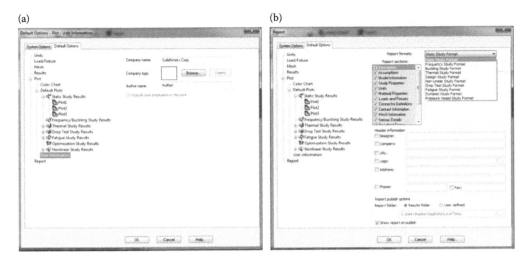

Figure 4.4
(a) User Information and (b) Report panels.

Table 4.2
Results Sections for Static Study

Report Sections	Items Related to Each Report Section by Default	Comments and Additional Data
Description		User-defined comments
Assumptions		User-defined comments
Model Information	• Names of all referenced components and sub-assemblies • Configuration name used in study • Path to all documents listed • Revision or last modified date • Physical properties of each body • Ply information for composites	User-defined comments
Study Properties	• Solution information including study name, analysis type and mesh type • Solver information including solver used in the study • Solver options including properties of the selected study • Result folder	User-defined comments
Units	Unit system for • Length/displacement • Temperature • Angular velocity • Stress/pressure	User-defined comments

(continued)

Table 4.2 (Continued)
Results Sections for Static Study

Report Sections	Items Related to Each Report Section by Default	Comments and Additional Data
Material Properties	• Material details such as name, yield strength, tensile strength • Values and graph of the material table • Descriptive information for custom-defined materials • Custom material properties defined by the user for the custom-defined materials	User-defined comments
Loads and Fixtures	• Fixture names, images, details and resultant forces • Load names, images and details • Values and graphs of the loading for transient analysis	User-defined comments
Connector Definitions	• Types of connectors and details • Connector forces	User-defined comments
Contact	• Types of contacts, images and properties	User-defined comments
Mesh Information	• Mesh details and image • Mesh control information (name, image and details)	User-defined comments
Sensor Results	• Sensor names, locations and details	User-defined comments
Resultant Forces	• Reaction forces • Reaction moments	User-defined comments
Beams	Select to include beam stresses or forces at • Joints • Ends • Extreme values • Entire length	User-defined comments Optional information: • Include beam forces • Include beam stresses
Study Results	• Includes images of all the plots in the Results folder • Clears or adds manually the desired images using insert image	User-defined comments Optional information: • Include images of all plots
Conclusion		User-defined comments
Appendix	Includes an Appendix with external data, either embedded or linked	User-defined comments Optional information: • Browse for embedded objects or links

All options selected through the **Options** property manager can be saved as they are any time by clicking the **OK** button or can be rejected by clicking **Cancel**. At any time, the **General options** can be set to their default values by clicking the **Reset All** button, displayed at the left bottom of the property manager (Figure 4.1b).

As far as the properties of the visualised results of our study are concerned, we will set the following options starting with opening of the **Options** property manager:

Simulation → Options (Figure 4.1a)

Thus, we open the **General System Options** window (Figure 4.1b). Then we

- Check both **What's Wrong messages** and **Shaded Load/Fixture symbol quality** – Shaded to be displayed
- Leave **Mesh colors** and **Result plots** options as they are by default and do not change the font properties (Figure 4.1c)
- Check the last two boxes to run the **Simulation Advisor and Command Manager**, hide the excluded components from the analysis and show the study material appearances.

Open the **Default Options** panel and set the options as given below:

- **Units** (Figure 4.2a) – Check **SI** (meter kilogram second) (1a, Figure 4.2a) and input the following units: Length/Displacement – mm; Temperature – Celsius; Angular velocity – rad/s; Pressure/Stress – N/m^2 (1b, Figure 4.2a).
- **Load/Fixture** (Figure 4.2b) – Leave the size (2a, Figure 4.2b) and the colour (2b, Figure 4.2b) of the symbols as they are by default. Check the preview option (2c, Figure 4.2b).
- **Mesh** (Figure 4.2c) – Set the quality of the mesh to High and **Jacobian points** to 4 (3a, Figure 4.2c). Choose Standard **Mesher Settings** (3b, Figure 4.2c) but do not check the other two automatic options. If necessary, they could be activated later through the **Mesh** property manager. Finally, you can either check the last two options or not (3c, Figure 4.2c). Their activation is not directly related to the current analysis, considering that the model is a solid body and there are no contact conditions to be defined.
- **Results** (Figure 4.2d) – Check the Automatic **Default solver** (4a, Figure 4.2d). Thus, the program will choose the solver itself comparing the **Direct sparse** to **FFEPlus**. Then we have to input the path to the directory where the data will be kept. We check the **SolidWorks** document folder, and the result files will be kept in the directory where your **SolidWorks** CAD files are located (4b, Figure 4.2d). Then we activate the **Trend Tracker** to keep the previous version in case of failure (4c, Figure 4.2d).
- **Plot** (Figure 4.3) – Leave all checks as they are by default for **Plot** (Figure 4.3a) and **Color Chart** (Figure 4.3b). If necessary, they can later be turned off or be activated for each separate plot. Neither **New Plots** nor **New Folders** will be added (Figure 4.3c). The number of plots for **Static Study** is preserved to be three, as it is by default (4a, Figure 4.3d). For **Plot1**, **Results type** is set to Nodal stress (4b, Figure 4.3d) and **Results component** to VON: von Mises stress (4c, Figure 4.3d). For **Plot2**, these options are Displacement and URES: Resultant Displacement, and for **Plot3**: Element Strain and ESTRN: Equivalent Strain.

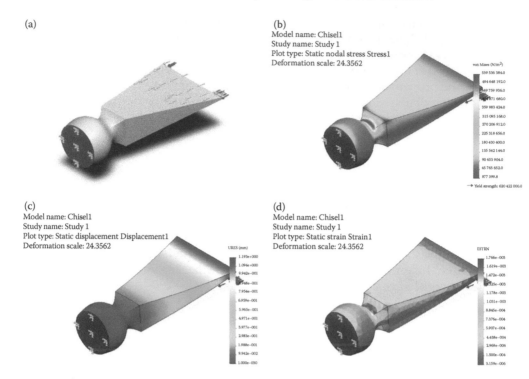

Figure 4.5
Some result plots of the static study of the chisel. (a) Chisel solid body model; (b) von Mises stress plot (N/m²); (c) Displacement plot (mm); (d) Element strain equivalent energy.

- **User Information** (Figure 4.4a) and **Report** (Figure 4.4b) – As this is a study case, there is no need for any specific data to be introduced here. Leave all options as they are by default.
- Click **OK** to keep all changes and to close the **Options** property manager.

The results of the analysis performed in Chapters 2 and 3, according to the above set options, are given in Figure 4.5.

We studied what and how result properties can be set before running the analysis. These properties affect the visualisation of the entire results data, yet they can be modified later if necessary.

We learned how to

- Define default General system options
- Define default analysis options, including units, loads and restraints, mesh settings and the default solver
- Set the default properties and types of plots
- View plots
- Generate automatically the final report
- Add user-defined information in the report
- Set the directories, where all result files and the report will be saved

4.2 DIFFERENT WAYS TO SYSTEMATISE AND PLOT THE RESULTS OF FEA

After running the study and reaching the post-processor stage, all results that are kept in the computer memory can be systematised and viewed. The program enables a few ways to present the results and thus to make their analysis easier. These ways are as follows:

- **Plotting the results** – probably the most commonly used way, especially in the early stages of the analysis. It makes it easier to get a general overview of the type of plotted results.
- **Listing the results** – enables numerical presentation of the results. It consists of enormous quantity of numbers, and sometimes, it is difficult to get the entire picture; yet this type of result presentation is preferred when a higher level of accuracy is needed.
- **Drawing graphs** – helps to see how a certain result type changes versus one parameter (for example, versus a geometrical position or time). It is a reasonably good way to combine visual and numerical data, especially throughout vulnerable zones.
- **Generating reports** – this is a very useful way of presenting the entire analysis data. The embedded structure of the automatically generated reports guides successfully the beginners throughout the presentation of their study.

All commands related to the presentation and systematisation of the results can be activated through either of the following:

- **Simulation Advisor** (⚑, 1c, Figure 4.6c), which can be started through the **SW Simulation command bar – Study Advisor** (🔍, 1a, Figure 4.6a)
- **Results Advisor** (📖, 1b, Figure 4.6b) and the corresponding pop-up menu
- **SW Simulation analysis tree** by right clicking on the **Results** folder (2a, Figure 4.6c) and picking any command of the pop-up menu (2b, Figure 4.6c).

The **Simulation Advisor/Results** guides the user through the display of the results (Figure 4.6d). As this is the easiest way of systematising the results, we will start our discussion about the display of the FEA results with it.

4.2.1 Results Display through Simulation Advisor

Probably this is the easiest way to systematise the results, because the user is guided step by step by the **Simulation Advisor**. If you answer correctly and cleverly to all questions that the **Simulation Advisor** (⚑) asks, you will have a reasonably good analysis of the final results in the end. This way of viewing the results is recommended for beginners.

There are three ways to activate the **Simulation Advisor**:

- Through the **SW Simulation command bar** by clicking the **Study Advisor** icon (🔍, 1a, Figure 4.6a), and after that, selecting the **Results Advisor** from the **Simulation Advisor** panel (3a, Figure 4.6d), which opens on the right side of the **Graphics area**

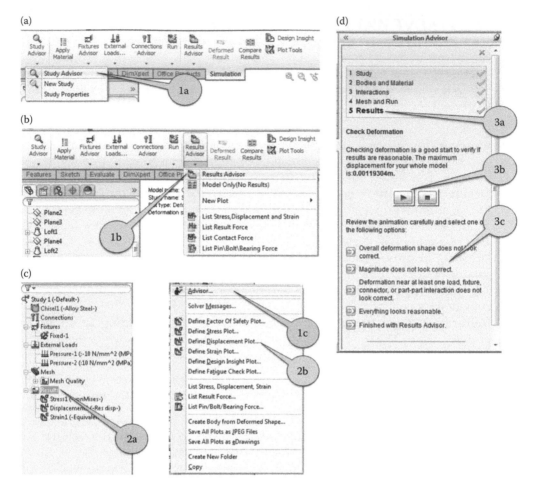

Figure 4.6
Activating Simulation Advisor – Results. (a) Through SW Simulation command bar – Study Advisor; (b) through SW Simulation command bar – Results Advisor; (c) through SW Simulation analysis tree – Results folder; (d) Simulation Advisor – Results Advisor.

- Through the **SW Simulation command bar** by clicking the **Results Advisor** icon (▣, 1b, Figure 4.6b), which directly opens the **Results Advisor** window of the **Simulation Advisor** (Figure 4.6d)
- Through the **SW Simulation analysis tree** by right clicking the **Results** folder (2a, Figure 4.6b) and then clicking on the **Advisor** icon at the newly opened pop-up menu (☞, 1c, Figure 4.6b)

The next step is to follow strictly the recommendations of the **Simulation Advisor** (☞). The systematisation of the results starts with

- **Checking the deformations** – By clicking the play button (3b, Figure 4.6d), the program generates a video clip to present the process of deformation of the body (Figure 4.7). Additionally, the value of the maximum displacement

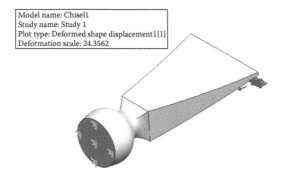

Model name: Chisel1
Study name: Study 1
Plot type: Deformed shape displacement1{1}
Deformation scale: 24.3562

Figure 4.7
Deformed shape of the chisel.

is given. In our case, it is 0.00119 m (see Figure 4.6d). After that, based on your experience, you have to assess the result in one of the following possible ways (3c, Figure 4.6d):

- Overall deformation shape does not look correct.
- Magnitude does not look correct.
- Deformation near at least one load, fixture, connector or part–part interaction does not look correct.
- Everything looks reasonable.
- Finished with Results Advisor.

Depending on your response, the program suggests some reasons for not achieving the correct deformed form or starts the next step.

- If **the generated deformed shape is wrong** (Figure 4.8a), the most common reasons according to the program are wrong load direction and amplitude. Possible suggestions for improvement include review of the loads and fixtures, re-examination of your expectations and, after finding the problem and correcting it, re-running the analysis.
- If **the magnitude of the generated deformed shape seems wrong** (Figure 4.8b), possible reasons could be as follows: the scale of the animation is not correct; the load is applied to each separate entity instead of the total; the model is under- or over-restrained; the material properties are not correct; the real deformations are nonlinear, and stress stiffening or softening occurs; and the load magnitude or its direction is wrong. It is very difficult to judge whether the model or your expectations are wrong. Hence, it is better to check the model carefully and, if everything seems correct, to re-examine your expectations.
- If **the displacements near loads, fixtures, connectors or contacts seem wrong** (Figure 4.8c), the most common reasons could be wrong type or settings of the fixtures, that is, the fixture restricts or allows movement that the physical model does not perform; load causes unexpected deformations; and parts are bonded when they should move relatively or just the opposite. The program's suggested possible ways to eliminate the problem include revision of the applied fixtures and loads or revision of the user's expectations.
- If **the stiffness of the model does not seem correct** (Figure 4.8d), the program suggests to use a stiffer material; if bending is a problem, to increase

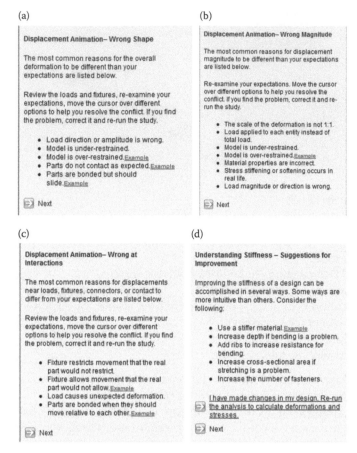

Figure 4.8
Possible causes of the observed displacement results. (a) Wrong Shape; (b) Wrong Magnitude; (c) Wrong at Interactions; (d) Understanding Stiffness.

depth or to add ribs to increase resistance to bending; and if stretching is a problem, to increase cross-sectional area or simply to increase the number of fasteners.

All these data are systematised in Table 4.3.

If the deformation movie and all corresponding data seem correct, by clicking the option **Everything looks reasonable**, the next step of viewing the results is activated (3c, Figure 4.6d).

This is the **Other Result Concerns** window (Figure 4.9), as different result types address different concerns, and the final goal is obtaining satisfying and accurate results. This window enables the consideration of some other possible reasons for not obtaining precise results (Table 4.4). Other options that can be selected are as follows (Figure 4.9a): Material breaking or yielding, Failure under repeated loading and unloading, Other failure modes or Finished with Results Advisor.

Regarding our choice, the result view continues with the following:

- If the selected option is **Material breaking or yielding**, plotting of the following results is enabled (Figure 4.9b):

Table 4.3
Possible Causes of the Observed Displacement Results

Possible Options after the Review of the Animation	Possible Reasons for the Observed Results	Possible Ways to Eliminate the Problem
Overall deformation shape does not look correct	• Wrong shape	• Load direction or amplitude is wrong. • Model is under-restrained. • Model is over-restrained. • Parts do not contact as expected. • Parts are bonded but should slide.
	• Wrong magnitude	• The scale of the deformation is not 1:1. • Load applied to each entity instead of total load. • Model is under-restrained. • Model is over-restrained. • Material properties are incorrect. • Stress stiffening or softening occurs in real life. • Load magnitude or direction is wrong.
	• Wrong at interactions	• Fixture restricts movement that the real part would not restrict. • Fixture allows movement that the real part would not allow. • Load causes unexpected deformation. • Parts are bonded when they should move relative to each other.
	• Understanding stiffness	• Use a stiffer material. • Increase depth if bending is a problem. • Add ribs to increase resistance for bending. • Increase cross-sectional area if stretching is a problem. • Increase the number of fasteners.
Magnitude does not look correct	• Wrong magnitude	• The scale of the deformation is not 1:1. • Load applied to each entity instead of total load. • Model is under-restrained. • Model is over-restrained. • Material properties are incorrect. • Stress stiffening or softening occurs in real life. • Load magnitude or direction is wrong.

(continued)

Table 4.3 (Continued)
Possible Causes of the Observed Displacement Results

Possible Options after the Review of the Animation	Possible Reasons for the Observed Results	Possible Ways to Eliminate the Problem
Magnitude does not look correct	• Wrong at interactions	• Fixture restricts movement that the real part would not restrict. • Fixture allows movement that the real part would not allow. • Load causes unexpected deformation. • Parts are bonded when they should move relative to each other.
	• Understanding stiffness	• Use a stiffer material. • Increase depth if bending is a problem. • Add ribs to increase resistance for bending. • Increase cross-sectional area if stretching is a problem. • Increase the number of fasteners.
Deformation near at least one load, fixture, connector or part–part interaction does not look correct	• Wrong at interactions	• Fixture restricts movement that the real part would not restrict. • Fixture allows movement that the real part would not allow. • Load causes unexpected deformation. • Parts are bonded when they should move relative to each other.
	• Understanding stiffness	• Use a stiffer material. • Increase depth if bending is a problem. • Add ribs to increase resistance for bending. • Increase cross-sectional area if stretching is a problem. • Increase the number of fasteners.
Everything looks reasonable	• Next	
Finished with Results Advisor	• Final	

• The first plot is that of **Factor of Safety (FoS)**. It assesses the safety of the design based on the maximal stresses, the material properties and the failure criteria. Values in the range (0, 1) indicate failure and outline vulnerable zones. In our case, FoS_{min} = 1.15 (Figure 4.10a). There are some techniques that can help to increase FoS_{min}. They either improve the strength or reduce the stress. You can think about considering an alternative material, adding or increasing fillet radii, adding ribs or gussets or increasing the number of fasteners. All techniques except the first concern the construction of the real part. By default, material change means higher modulus of elasticity to increase the stiffness. Usually, the relation is proportional,

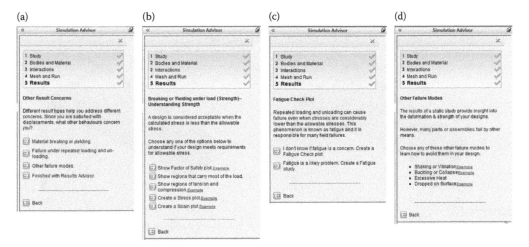

Figure 4.9
Other results concerns. (a) Other Result Concerns panel; (b) Understanding Strength; (c) Fatigue Check; (d) Other Failure Modes.

Table 4.4

Other Possible Causes of the Observed Displacement Results

Possible Options after Assessment of the Animation	Possible Reasons for the Observed Results	Possible Next Choices
Material breaking or yielding	• Breaking or yielding under load (strength) – Understanding strength	• Show Factor of Safety plot • Show regions that carry most of the load • Show regions of tension and compression • Create a Stress plot • Create a Strain plot
Failure under related loading and unloading	• Fatigue Check Plot	• I don't know if fatigue is a concern. Create a Fatigue Check plot • Fatigue is likely a problem. Create a fatigue study
Other failure modes	• Other Failure Modes	• Shaking or Vibration • Buckling or Collapse • Excessive Heat • Dropped on Surface
Finished with Results Advisor	• Final	

and the higher the modulus, the less the deflection. Another very important question is whether the results are accurate. You must remember that sometimes the displacements can seem accurate enough, while the stresses can bother you. The reason is the workflow of the used calculation techniques; that is, first, the displacements are calculated and, after that as a chain, based on strain relations and Hook's law, the stresses. A very simple way to overcome that uncertainty about the stress results is to

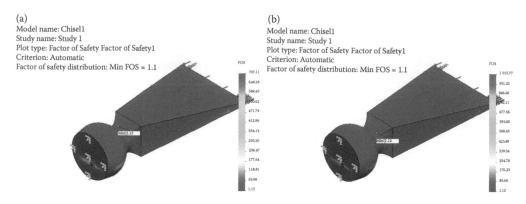

Figure 4.10

FoS plots. (a) Using none-adaptive method; (b) using h-adaptive method.

generate a finer mesh and to re-run the study. If the results do not change significantly, everything is OK. Otherwise, you can use automatic adaptive methods or enable the h-adaptive method and re-run the study. In our case, just to compare the FoS results for non-adaptive and h-adaptive methods, the analysis is re-run and the result is that the h-adaptive method has satisfied the current accuracy of 98.3133%. There is no need to increase the target accuracy $\text{FoS}_{min} = 1.12$. Thus, $\text{FOS}_{min}^{\text{none-adaptive method}} > \text{FOS}_{min}^{\text{h-adaptive method}}$.

- The next step is to **Show regions that carry most of the load**. Thus, the **Design Inside** plot is started. It shows the volumes that work the hardest (Figure 4.11) and helps us to improve FoS levels. For example, if FoS is too low, additional material could be added in the vulnerable zones, or if FoS is unreasonably high, the body weight can be reduced by removing some material. This plot guides us through these constructional improvements by showing different load levels all over the body. As a result, the body will have better construction and lower weight, and probably its final cost would be reduced as well.
- The next plot is **Shows regions of tension and compression**. It plots areas with tensile stresses versus compressed areas. There are two really large groups of materials. The first group includes the **ductile materials**,

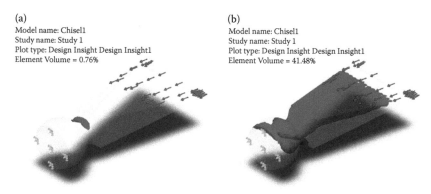

Figure 4.11

Plots of the regions that carry most of the loads. (a) Regions that are loaded more than 90%; (b) regions that are loaded more than 30%.

which carry well-enough tension and compression. Almost all alloys, steel, iron and many other materials belong to this group. The second group combines the **brittle materials**, such as concrete, cast iron, etc. They work very well when exposed to compression, but unfortunately, their reliability, when exposed to tension, is questionable. That is why no matter to which group the material of the body belongs, tensile stress is more of a concern than the compressive one, especially when yielding or fracture is a concern. Additionally, when the body is bended, based on our experience, it is reasonable for tensile stresses to be spread on one side of the body (the side that is opposite of the entity at which the load is applied) and the compressive stresses on the other (Figure 4.12).

- **Stress plot** presents the distribution of different stresses over the body contour or inside. In our case, the von Mises stress plot is given (Figure 4.13a). Its maximum is about 540 MPa in an area at the gudgeon of the chisel. The yield stress for the selected material is 620 MPa. Therefore, there is no need for any constructive changes regarding the appearance of the chisel to be made.
- The last proposed by the **Results Advisor** plot is the **Strain plot**. It is used to gain an idea of the inside of the body strain. This plot is useful if there are any on-site strain measurements. The existing relation between

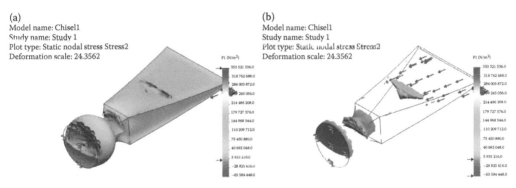

Figure 4.12
Plots of regions of tension and compression. (a) Regions of tension; (b) regions of compression.

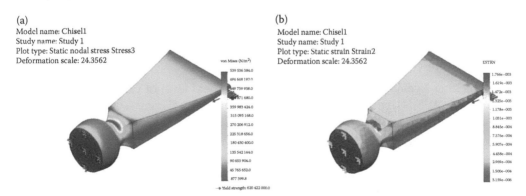

Figure 4.13
Plots of stress and strain. (a) von Mises plot; (b) Equivalent strain plot.

strain and stresses helps for easier assessment and comparison of the accuracy of these plots. The **Equivalent strain** plot is shown in Figure 4.13b. The maximal values are at the same area as the maximal von Mises stresses.

- The next possible option deals with **fatigue phenomenon** (Figure 4.9c). This is a process during which stresses less than the allowable can cause failure due to their constant variation, particularly by repeated loading and unloading of the body. This phenomenon is known as fatigue, and it is the reason for many unpredicted field failures, especially in structures made of brittle materials like cast iron. There are two possible answers to the question regarding fatigue. The first is to create a **Fatigue Check** plot if we are not aware of the influence of the fatigue. The second choice is to assume that fatigue might be a problem and to develop **Fatigue study** (🔁).

- The last option has to do with **Other Failure Modes** (Figure 4.9d). Usually this option is not directly related to **Static study** and suggests possible reasons for failure. They can be observed during dynamic loading such as Shaking and Vibration; can be due to buckling phenomenon such as Buckling or Collapse; can be related to thermal effects such as Excessive Heat; or finally can be a consequence of dropping of the body such as Dropped on Surface.

In conclusion, it must be admitted that viewing the results through the **Results Advisor** strongly limits their presentation. As you have probably noticed, only certain plots have been shown, and there are no accompanying lists or graphs. That is why the other two ways of viewing the results are more commonly used.

We learned that viewing the results can be done in different ways and one of them uses the **Results Advisor** (Figure 4.6d).

It starts with a **Check of the deformations** (Figure 4.7), where in order to continue, we are expected to select one of the suggested answers. It is possible for

- The generated **deformed shape to be wrong** (Figure 4.8a)
- The **magnitude of the generated deformed shape to seem wrong** (Figure 4.8b)
- The **displacements near loads, fixtures, connectors or contacts to seem wrong** (Figure 4.8c)
- The **stiffness of the model not to seem correct** (Figure 4.8d)

There are different ways to eliminate the problems, if there are any. The most common of them are prompted by the **Results Advisor**; others depend on our knowledge and experience.

If clicking **Everything looks reasonable**, the next level of viewing the results is reached. At this point, some **Other Result Concerns** (Figure 4.9a) occur. They can be

- **Material breaking or yielding** (Figure 4.9b). To assess the results correctly, some plots can be generated. They are the **FoS** plot (Figure 4.10); the **Show regions that carry most of the load** or **Design Inside** plot (Figure 4.11); the **Show regions of tension and compression** plot (Figure 4.12); and the **Stress** plot (Figure 4.13a) and **Strain** plot (Figure 4.13b).

- **Fatigue phenomenon** (Figure 4.9c), where there are two available options: to create a **Fatigue Check** plot or to develop **Fatigue study**.
- **Other Failure Modes** (Figure 4.9d), including Shaking and Vibration; Buckling or Collapse; and Excessive Heat or Dropped on Surface.

Finally, we click the **Finished with Results Advisor** to close the **Simulation Advisor** (Figure 4.6d).

During this chapter, guided by the Results Advisor, we learned how to

- Generate deformed shape clips and plots
- Check for possible reasons for incorrect deformations and consider the most common ways to eliminate the problem
- Consider some Other Result Concerns, related to stress and strain distribution
- Make some changes in the design to improve the final results

4.2.2 Results Display through Results Folder in the Analysis Tree

Another way to display the results is through the **Results** folder in the analysis tree (Figure 4.6c). You can start viewing the results by either showing the plots that are at the analysis tree (Figure 4.14a), and which number and properties are defined by default (see Section 4.1), or by creating new plots, listings and drawings using the **Results** pop-up menu (Figure 4.14b), which can be opened by right clicking on the **Results** in the **SW Simulation analysis tree.**

This time, the view of the results will start with the default plots (Figure 4.14a). If we right click on the plot that is not highlighted/active, a small pop-up **Property** menu opens (Figure 4.15a). It includes only a few of the commands that are accessible in the larger **Property** menu, which opens after right clicking on an active plot (Figure 4.15b).

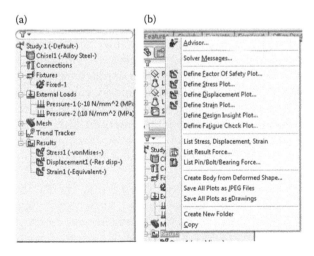

Figure 4.14
Display of the results through the Results folder in the analysis tree. (a) Analysis tree with default plots; (b) Results pop-up menu.

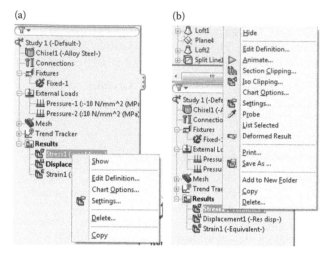

Figure 4.15
Property menus of plots. (a) Pop-up Property menu of a non-active plot; (b) pop-up Property menu of an active plot.

Each plot can be selected either by double left clicking on its name in the **SW Simulation analysis tree** or by selecting the **Show** command in the pop-up **Property** menu (Figure 4.15a). After selecting the plot, it is automatically displayed in the **Graphics area**. Its features coincide with the plot's pre-defined properties.

All these properties can be changed through the pop-up **Property** menu (Figure 4.15b). The commands from that menu are discussed in detail below:

- The **first** command is an alternative **Show/Hide** command. It either activates or deactivates the current plot.

The **second group** combines commands through which the user can directly influence the properties of the plot visualisation.

- **Edit Definition** – This command is available in the two **Property** menus (Figure 4.15). At first, we will explain how this command works when it is associated with the **Stress** plot, and after that, we will compare it to the corresponding **Edit Definitions**, related to **Displacement** and **Strain** plots. Clicking on **Edit Definition** opens the window shown in Figure 4.16a. All options related to that command are grouped in **Display** (1b, Figure 4.16a), **Advanced Options** (1c, Figure 4.16a), **Deformed Shape** (1d, Figure 4.16a) and **Property** (1e, Figure 4.16a) sub-windows. It is important to remember the operation of the **OK** (✔), **Cancel** (✖) and **Pin** (📌) functions (1a, Figure 4.16a).
  - Through the **Display** sub-window, you have to choose the **Component** (📊, Figure 4.16b) to be displayed and the corresponding **Units** (📏, Figure 4.16c). There are 13 stress components accessible for solid body models and you have to select one of them. These components are combined in a few groups: **Normal stresses – SX**: X normal stress; **SY**: Y normal stress; **SZ**: Z normal stress (2a, Figure 4.16b); **Shear/Tangential stresses – TXY**: shear in Y direction on the YZ plane; **TXZ**: shear in Z direction on the YZ plane; **TYZ**: shear in Z direction on the XZ plane (2b, Figure 4.16b);

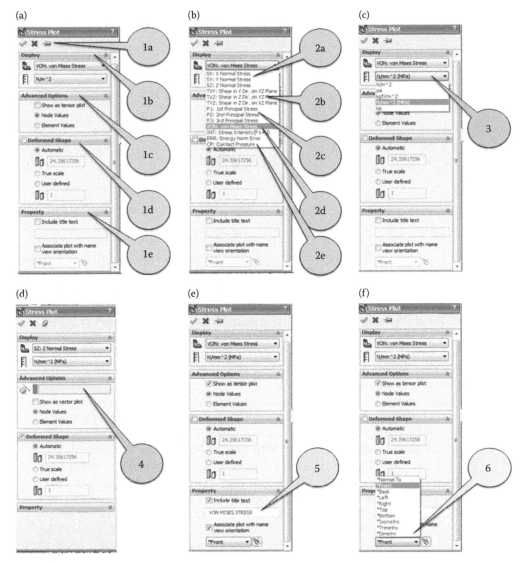

Figure 4.16
Edit Definition property manager of Stress plot. (a) Edit Definition for Stress plot; (b) Display menu; (c) units of the selected parameter; (d) input of reference entity (optional); (e) input of title; (f) associate plot with name view orientation.

Principal stresses – P1: 1st principal stress; **P2**: 2nd principal stress; **P3**: 3rd principle stress (2c, Figure 4.16b) and **von Mises Stress** (2d, Figure 4.16b), **Stress Intensity** (P1–P3), **ERR**: Energy Norm Error, **CP**: Contact Pressure (2e, Figure 4.16b). Directions of the stresses are based on the selected reference geometry or on the original orthogonal coordinate system, which is set by default. Here, I would like to remind you that the stress state in one point of the solid body can be completely defined by 18 different stresses – 3 at each side of the stress cube (Figure 4.17a) but only 6 of them are independent. Thus, to describe the stress state in a point, it is enough to know the values of the 6 independent stresses,

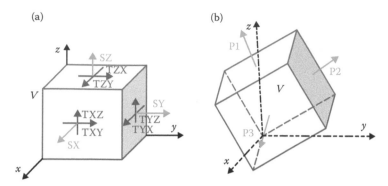

Figure 4.17
Stress cube. (a) Stress cube with basic stresses; (b) stress cube with principal stresses.

which include 3 normal (SX, SY, SZ) and 3 shear stresses (TXY, TXZ, TYZ). If the stress cube rotates about the point, the values of the stresses change. It is always possible to orientate the cube in such a way that there will be no shear stresses in its sides. Then the normal stresses will coincide with the principle stresses (P1, P2, P3, Figure 4.17b). Plots of **First Principal Stress** (P1, σ_1) and of **Tangential Stress** (TYZ, τ_{YZ}) are shown in Figure 4.18a and b. The next two plots show dependency of the plotted component on the reference geometry. The plot of **Normal Stress** SZ (SZ, σ_Z, Figure 4.18c) uses the initial coordinate system, whereas the plot in Figure 4.18d uses a newly defined coordinate system, in which the Z axis is parallel to the X axis of the initial system. Thus, the stress distribution in this plot coincides with the SX plot using the initial coordinate system.

- The second sub-window of the **Edit Definition** property manager combines **Advanced Options** (Figure 4.16d). The first option is the **Reference geometry** (⬡, 4, Figure 4.16d). The reference plane, axis or coordinate system can be selected through clicking on the signature in the floating **Design tree** in the **Graphics area** or by directly clicking on the entity. The signature of the selected entity is automatically displayed in the blue window. This option is accessible only for directional stresses, such as normal or shear stresses. In spite of its availability, the input of reference entity is not mandatory, and the program uses the original coordinate system by default. The second option controls the visualisation of the stresses: chart plot (Figure 4.18e), vector or tensor plot (Figure 4.17f). The tensor plot of von Mises stresses is shown. The vector mode plots the stresses at each node in relation to their magnitude and direction. After that, either a node mode or an element mode must be selected. The node mode (**Node Values**, Figure 4.16d) generates plots based on the calculated stress component values in the nodes and inner interpolation among them. As a rule, the plot is smooth and good-looking (Figure 4.18). The element mode (**Element Values**, Figure 4.16d) generates plots based on the calculated stress components at the centres of the elements – one value/colour for each element. This mode is considered a criterion about the quality of the mesh. If the plot is rough and the difference in colours of two neighbouring

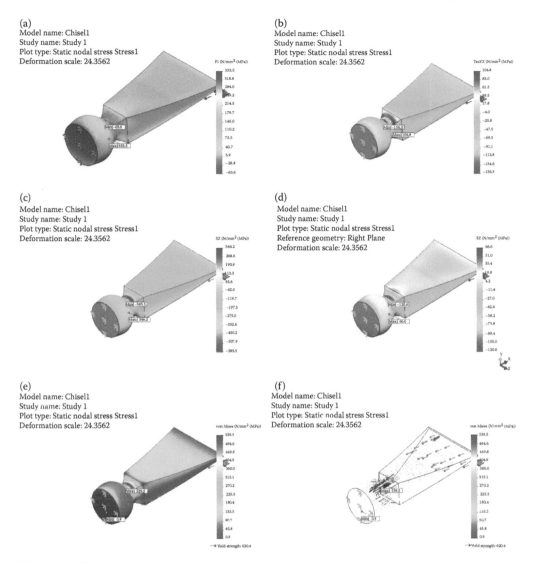

Figure 4.18
Different Stress plots. (a) Plot of First Principal Stress (P1, σ_1); (b) plot of Tangential Stress (TYZ, τ_{YZ}); (c) plot of Normal Stress SZ (SZ, σ_Z); (d) plot of Normal Stress SZ with a reference plane (SZ, σ_Z); (e) plot of von Mises Stress (VON, σ_{red}); (f) Tensor plot of von Mises Stress (VON, σ_{red}).

elements seems unreasonably contrasting, then the mesh is too coarse and the model should be re-meshed with a finer mesh and re-run. If the plot is smooth and good-looking, the quality of the mesh is OK.

- The next displayed sub-window defines the properties of the deformed shape. The existing possible options are as follows: **Automatic – Scale Factor** (⬛⬛) is automatically calculated in a way that the program scales the largest deformation to 10% of the largest dimension of the smallest box that surrounds the model; **True scale** – Scale Factor = 1; **User defined** – the user inputs the scale factor.

- The last sub-window is titled **Properties** (Figure 4.16e and f). It enables the input of a title text to the plot and the association of the current view with the active plot – **Associate with current view** (⬚), regarding or not the **Zoom to fit** function.
- **Animate** (▷, Figure 4.15b) – This command is available only in the pop-up **Property manager** of an active plot. For a static study, this option creates a video clip (*.avi), which simulates the deformation of the body as a process. There are two sub-windows: **Basics** (1a, Figure 4.19) and **Save as AVI file** (1b, Figure 4.19).
  - The available **Basics** icons correspond to the common, well-known labels: **Play** (▶); **Pause** (‖); **Stop** (■); **Frames** (⬛), which sets the number of frames to be used in an animation; **Speed** (⬛), which controls the speed of the animation; **Forward Only** (➡), which plays the animation forward one time; **Loop** (⟳), which plays the animation in a continuous looping pattern; and **Reciprocate** (⬌), which plays the animation from start to end, then end to start, and continues repeating.

    The following are accessible through the second sub-panel commands: **Option**, which sets the compressor to be used, and **Browse** and optional **View with Media Player**.
- **Section Clipping** (▥, Figure 4.15b) – This command is available only in the pop-up menu of an active plot, just like the **Animation** command. It enables plotting different section views of the displayed result. It combines two or more sub-windows, each corresponding to a different section (Figure 4.20).
  - The first sub-window is titled **Section 1** (1, Figure 4.20a). By selecting one of the icons **Plane** (▧), **Cylinder** (▦) or **Sphere** (◯) (1a, Figure 4.20a), and pointing an appropriate reference entity directly in the **Graphics area** (1b, Figure 4.20a), different types of sections can be generated. By clicking the button **Reverse clipping direction** (⬙), which is at the left of the window, the direction of the cut can be flipped. Through the **Distance**

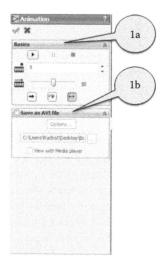

Figure 4.19
Animation property manager.

(a)   (b)   (c)

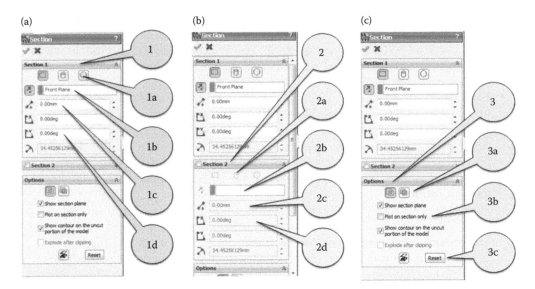

Figure 4.20
Section property manager. (a) Section 1 sub-window; (b) Section 2 sub-window; (c) Options sub-window.

icon (✎), the offset distance measured from the selected entity can be set (1c, Figure 4.20a). It uses the default units. By introducing some values in the next two windows – **Rotation X** (⌐) and **Rotation Y** (⌐) (1d, Figure 4.20a) – the section can be rotated around the **X** or **Y** axis. The last window – **Radius** (⌐) – is highlighted only when cylindrical or spherical sections are selected.

- The next sub-window is titled **Section 2** (2, Figure 4.20b). The available icons are the same as the ones in **Section 1**; thus, the numbers in the figure are identical. The **Section 2** sub-window is activated by checking the square on the left of the title. It adds a new section to the plot. Optionally, the **Section 3** sub-window becomes available after activating **Section 2** and so on. Defining more than one section allows viewing complex sections. The **Section** property manager enables plotting up to six sections simultaneously.

- The last sub-window is **Options** (3, Figure 4.20c). The two icons at the top of this sub-window are **Union** (▦) and **Intersection** (▦) (3a, Figure 4.20c). The first command combines the areas of all sections, whereas the second one displays only the common areas. The next few options (3b, Figure 4.20c) are related to the display of different entities in the plot. They are **Show section plane**, **Plot on section only**, **Show contour on the uncut portion of the model** (which is not active when **Plot on section only** is selected) and **Explode after clipping**.

- At the bottom of the sub-windows are the **Clipping on/off** (▦) icon and the **Reset** button (3c, Figure 4.20c), which sets the plot options to their initial state.

The **Section** options are confirmed and the plot is generated by clicking **OK** (✓).

Section plots allow us to plot the stress or another component distribution inside the solid body and thus help us to analyse the results. A simple (one entity) section and some combined (of two or more entities) ones are shown in Figure 4.21. They present the effect of different combinations of display options on the final plot.

- **Iso Clipping** (⬛, Figure 4.15b) – This command is available only in the pop-up menu of an active plot, just like the two previous commands. **Iso Clipping** views surfaces of a specified value or surfaces where the values are within a certain range (Figure 4.23c). Up to six different iso-surfaces can be defined (Figure 4.23d). There are few sub-windows in the **Iso Clipping** property manager (Figure 4.22). All **Iso** sub-windows (1, Figure 4.22) enable defining the plot properties.
  - There is a window for the input of the value of the iso-surface (1a, Figure 4.22); a slider for coarse adjustment of that value (1b, Figure 4.22); and a **Reverse clipping direction** icon (⬛, 1c, Figure 4.22) for flipping the cut. If this icon is active, values smaller than the input are shown (Figure 4.23a); otherwise, only surfaces where the values of the plotted property are larger than the input are presented (Figure 4.23b).
  - **Options** sub-window (2, Figure 4.22) differs a little bit from the corresponding one in the **Section** property manager. The new option is **Plot**

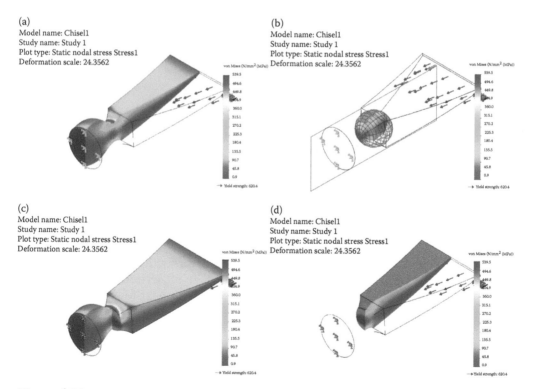

Figure 4.21

Sectioned stress plots. (a) Right plane section with shown contour on the uncut portion of the model; (b) intersecting right plane and sphere sections with shown section entities; (c) union of two planar sections with shown contour on the uncut portion of the model; (d) intersection of two planar sections with no shown contour on the uncut portion of the model.

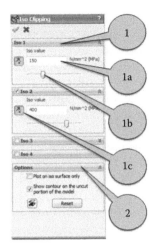

Figure 4.22
Iso Clipping property manager.

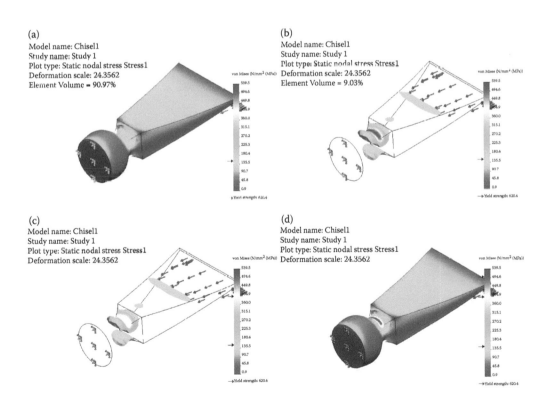

Figure 4.23
Iso Clipping plots. (a) Iso surface with stress values smaller than 150 MPa; (b) Iso surface with stress values larger than 150 MPa; (c) Iso surface with stress values in between 150 and 400 MPa; (d) Iso surfaces with stress values in between 150 and 400 MPa (yellow-red surface) and smaller than 150 MPa (blue surface).

on iso surface only. If it is selected, the program displays only the surface; otherwise, it displays surfaces, including the iso-surface, that have values larger or smaller than the specified value.

- **Chart Options** (Figure 4.15) – This is the second command accessible through both pop-up menus. It controls the display of the legend in the plot. There are three sub-windows in the **Chart Options** property manager: **Display Options**, **Position/Format** and **Color Options** (Figure 4.24a).
  - The **Display Options** sub-window includes four check buttons. The first two of them are **Show min annotation** and **Show max annotation** (1a, Figure 4.24b), which enable either displaying or not the minimum and the maximum values of the plot directly on the plotted **Graphics area** model. The model name, study name, the plot type and the deformation scale are displayed by checking the **Show plot details** (1b, Figure 4.24b). The last

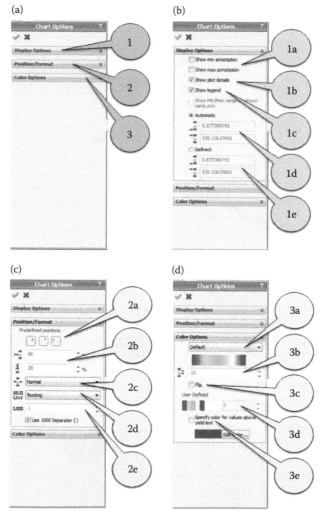

Figure 4.24
Chart Options property manager. (a) Chart Options window; (b) Default Options sub-window; (c) Position/Format sub-window; (d) Color Options sub-window.

check button is **Show legend** (1c, Figure 4.24b) and controls either the display or not of the legend. As the final check, the button **Show Min/ Max range on shown parts only** is active for assembled models only; it will not be discussed here. If **Automatic** is selected, the minimum (**Min**,) and the maximum (**Max**,) values of the chart automatically define its range (1d, Figure 4.24b). If **Defined** is selected, the minimum (**Min**,) and the maximum (**Max**,) values of the chart are specified manually (1e, Figure 4.24b).

- Next is the **Position/Format** sub-window (Figure 4.24c). The first buttons (2a, Figure 4.24c) specify **three predefined positions of the legend** in the plot – (at the right bottom), (right, in the middle) and (at the left bottom). The next two options are alternative options to set the legend's disposition in the plot (2b, Figure 4.24c). The first one ( ) defines **the horizontal distance** from the left side of the graphics area in percentage of its width, while the second one ( ) sets **the vertical disposition** of the legend. Optionally, if the **Chart Options** property manager is active, the legend can be positioned by right clicking on it and dragging it. This method for positioning the legend is easier, but sometimes it is very hard to displace the legend on one and the same position in different plots, relying only on the precision of our eyes and hands. The last icon on that sub-window (, 2c, Figure 4.24c) sets **the thickness of the legend bar**. The last icons help in defining the type of the display of the numbers, setting **the number format** by choosing among scientific, floating or general types (, 2d, Figure 4.24c); **the number of decimal places**, which can be up to 16 (, 2e, Figure 4.24c); and **the use of 1000 separator** or the use of different format for smaller numbers.

- **Color Options** is the third sub-window (Figure 4.24d). It enables choosing the colour palette or defining a new one (3a, Figure 4.24d). We can choose among Default ( ), Rainbow ( ), Gray Scale ( ) or User Defined palette. After that, the number of colours used in the chart should be set (, 3b, Figure 4.24d); the direction of colour mapping can be preserved or flipped (3c, Figure 4.24d), or the User Defined palette (3d, Figure 4.24d) is defined. It is based on up to nine user-defined colours, but no colour interpolation and shading are enabled, that is, only the user-defined colours are used. Finally, the zones, where the von Mises stresses are larger than the yield strength, can cither be coloured or not (3e, Figure 4.24d).

Almost the same **Position/Format** and **Chart Options** have been defined through the path Simulation → Options → Default Options → Plot → Color Chart.

The main difference between these two ways is that the options defined through the **Options** window (Figure 4.3a and b) set the properties of all plots; the options defined through the **Chart Options** property manager affect only the active plot.

The effect of different **Chart Options** combinations is shown in Figure 4.25.

- **Settings** (, Figure 4.15) – This is the last command in the second group, which is accessible through pop-up menus of the active and non-active plots. It controls the setting of the plot, particularly **Fringe**, **Boundary** and **Deformed Plot** options (Figure 4.26a).

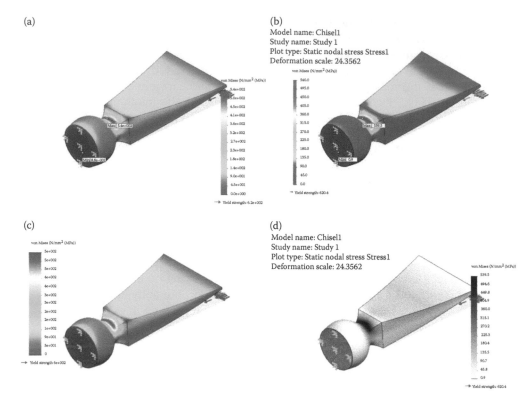

Figure 4.25
von Mises stress plot with different chart options. (a) Displayed min/max values; Chart – at the right bottom; Scientific type of numbers; Default palette. (b) Displayed plot details; Chart – at the left bottom; Narrow chart; Floating type of numbers; Rainbow palette. (c) No min/max values; no plot details; Chart – at the right top; Thick chart; General type of numbers; Default palette. (d) No min/max values; Chart – at the left middle; Floating type of numbers; Gray palette.

- The first sub-window sets **Fringe** options (Figure 4.26b). The choice is among **Point** (coloured point contours), **Line** (coloured line contours) and the colour-filled display either with discrete shading (**Discrete**) or with smooth shading (**Continuous**). The most commonly used display among the above mentioned is **Continuous**.
- The second sub-window sets **Boundary Options** (Figure 4.26c). There could be no boundary edges (**None**), or the plot can be superimposed on the model with displayed boundary edges (**Model**) or be superimposed on the plotted mesh (**Mesh**).
- Finally, we must set the **Deformed Plot Options** (Figure 4.26d). The plot of the deformed shape can or cannot be superimposed on the undeformed shape. The settings of the undeformed shape, particularly its colour and transparency level, are set through that sub-window.

The effect of different **Settings** combinations is shown in Figure 4.27.

- The next, available only in the pop-up menu of an active plot command, is the **Probe** command (🖉, Figure 4.15b). It helps displaying the numerical value at a picked point, drawing a graph and viewing the trend or simply

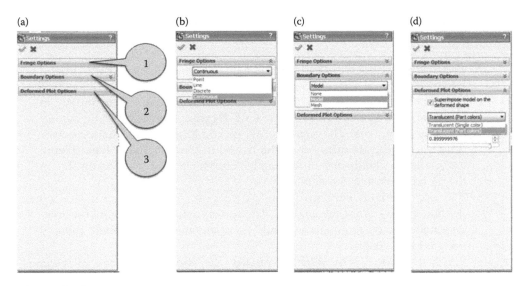

Figure 4.26
Settings property manager. (a) Settings window; (b) Fringe Options sub-window; (c) Boundary Options sub-window; (d) Deformed Plot Options sub-window.

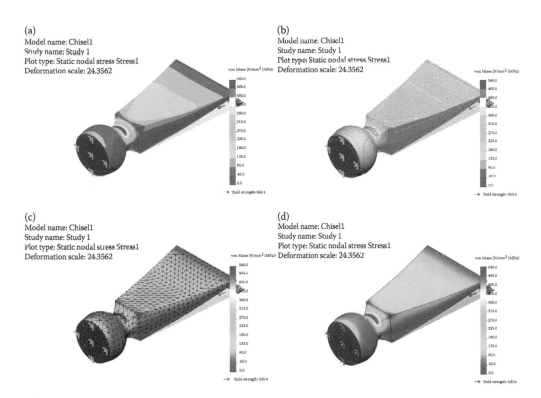

Figure 4.27
von Mises stress plot with different settings. (a) Discrete fringe; no boundary outline; blue unde-formed shape. (b) Line fringe; no boundary outline; grey undeformed shape. (c) Continuous fringe; mesh outline; grey undeformed shape; transparency level – 80%. (d) Continuous fringe; model outline; grey undeformed shape; transparency level – 60%.

saving some of the results. According to the selected **Options – At location** or **On selected entities**, there are four or five sub-windows in the **Probe** property manager.

- The first sub-window is **Options** and it enables the selection of the probed parameter (Figure 4.28b). The first option is **At location**. When this option is selected, the next step is to pick a vertex or a node either directly in the **Graphics area** or from the floating **Design tree** at the right side of the **Graphics area**. The results at the picked location are simultaneously displayed in the **Results** sub-window and in the **Graphics area** (Figure 4.28b and c). The displayed X, Y and Z coordinates are the coordinates in the global coordinate system.

- The option **From sensors** is not active now, for no sensor has been defined at the start of the analysis. The sensor definition is useful for time-dependent or design analysis, while for static analysis, generally there is no need for sensors.

- The last option is **On selected entity** (🗊). We have chosen to select an edge (Figure 4.28f), but these can also be faces, edges, etc. To display the results, the **Update** button must be clicked (Figure 4.28d). The data for all nodes/elements related to the picked entity are displayed in the **Results** sub-window (Figure 4.28d). Those included in the table properties are defined in the **Annotation** sub-window (Figure 4.28e). When an entity is selected, the **Summary** sub-window is active. It provides some statistics on the data of the selected entity, including the sum (**Sum**), the average value (**Avg**), the **Max** and **Min** values and the **RMS** (root mean square) value. It must be remembered that in case of n values

$$x_i, X_{RMS} = \sqrt{\frac{1}{n}(x_1^2 + x_2^2 + \cdots + x_n^2)} \cdot$$

- The **Report Options** sub-window (Figure 4.28g) provides some options of saving the results. The first option is **Save as Sensor** (🖼) and is accessible when a sensor is defined. The second option is **Save** (🖼). It saves the probe results in a file of type *.csv, which opens either with a text editor or with MS Excel. By clicking the **Plot** icon (🖾), a graph of the results can be displayed. The generated graph can be saved either as a *.csv file or as a picture file. The options of the graph display are directly set for each separate picture. The X-axis displays a parametric presentation of the nodes in a range [0,1] (Figure 4.28h).

- The **List Selected** command (Figure 4.15b) operates in a similar way as the **Probe** command. Consequently, it will not be discussed in detail.

- **Deformed Result** (🗩, Figure 4.15b) displays the deformed shape of the analysed object.

Before finishing the discussion on the display of the active plot, some words about the **Standard View toolbar** (Figure 4.29) must be said. It is situated at the top of the **Graphics area**. It functions in one and the same way in **SolidWorks** as in **SW Simulation** and helps to orient the model in one of the presented standard views and to manage one or more plots. As far as the standard views are concerned, the icons shown in Figure 4.29 are used.

They are used to preview four different views of the von Mises stress plot in the **Graphics area** (Figure 4.30).

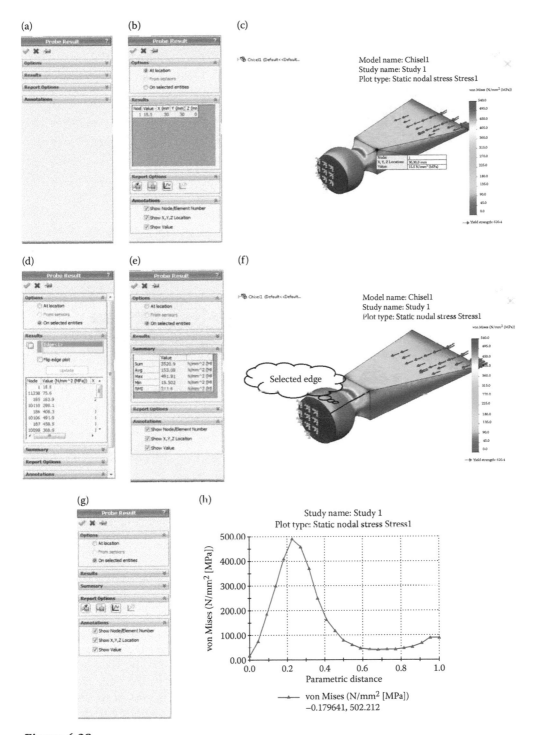

Figure 4.28
Probe property manager. (a) Probe Results window; (b) results at picked location; (c) von Mises stress at picked location; (d) results at selected edge; (e) summary of the results at selected edge; (f) edge picked at the Graphics area; (g) Report Options; (h) graph of von Mises stresses along the selected entity.

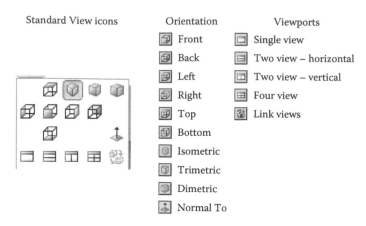

Figure 4.29
Standard View toolbar.

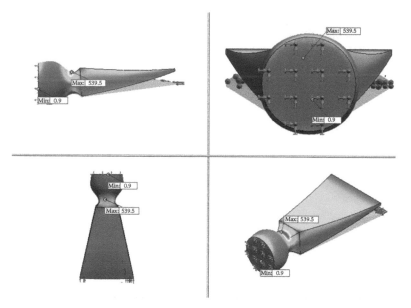

Figure 4.30
Four standard viewports of von Mises plots.

The last few commands included in the pop-up menu of the active plot (Figure 4.15b) are **Print**, **Save as** (⬚), **Add to New Folder**, **Copy** and **Delete**. As they are commonly used and familiar commands, their function will not be discussed here. The most specific for them is that they can be related either to the very plot in the **Graphics area** or to some generated graphs, to some lists of results or to results as a whole. The only advice that can be provided here as far as the use of these commands is concerned is to read and to answer carefully all questions set to you by the software, relying on your experience to work with computers.

If we are systematising displacements, both pop-up menus are the same as the ones shown in Figure 4.15. The commands and their options that are accessible through different windows are similar except for those given in the following.

- **Edit Definition** – This command is available in both pop-up menus (Figure 4.15). The first to be discussed is the **Display** sub-window (Figure 4.31a). All available components (🖦) are combined in two groups: **Displacements** – X Displacement (**UX**), Y Displacement (**UY**), Z Displacement (**UZ**) and Resultant Displacement (**URES**); and **Reaction Forces** – X Reaction Force (**RFX**), Y Reaction Force (**RFY**), Z Reaction Force (**RFZ**) and Resultant Reaction Force (**RFRES**) (Figure 4.31b). The directions of the displacements and the reaction forces are parallel to the axes of the global coordinate system. Some displacement plots are given in Figure 4.31c.
- Next, a little bit different command in the pop-up menu is the **Probe** command (🖉, Figure 4.15b). When the active displacement plot shows resultant displacements, a new option in the **Options** sub-window (Figure 4.32b) is accessible. This is the **Distance** button. It enables measuring the distance between any two nodes, picked in the **Graphics area** (Figure 4.32c). The distance after the deformation can be compared to its initial value (Figure 4.32a).

The **Distance** option helps in deformation analysis. The results of the compared distances for the undeformed shape and for the deformed chisel are given in Table 4.5.

The last of all default plots is the **Strain** plot. There are no differences in all applicable commands, except **Edit Definition**. The only differences, as it is reasonable to be expected, are the components (🖦) in the **Display** sub-window. They are systematised in four groups: **Normal strain** (**EPSX**: X normal strain; **EPSY**: Y normal strain; **EPSZ**: Z normal strain), **Shear strain** (**GMXY**: shear in the Y direction on the YZ plane; **GMXZ**: shear in the Z direction on the YZ plane; **GMYZ**: shear in the Z direction on the XZ plane), **Principal strain E1**: normal strain (1st principal direction), **E2**: normal strain (2nd principal direction), **E3**: normal strain (3rd principal direction), **Strain – total** (**ESTRN**: equivalent strain; and **SEDENS**: strain energy density and **ENERGY**: total strain energy, which are available only in element mode).

Before continuing our explanation, it will be useful to provide some information about what strain is and about the different types of strain. Just like stresses, strains are

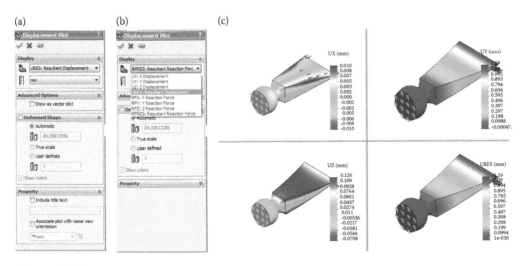

Figure 4.31
Setting displacement properties and displacement plots. (a) Edit Definition panel; (b) Display sub-window; (c) different displacement plots.

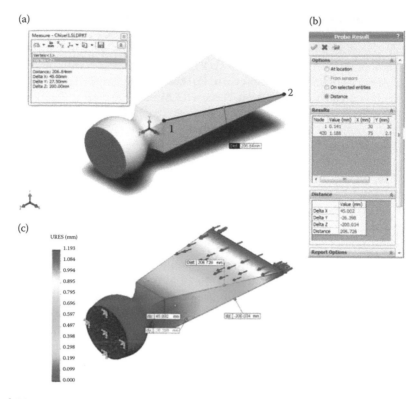

Figure 4.32
Distance calculations in deformed and undeformed chisels. (a) Distance at the undeformed chisel; (b) Options sub-window; (c) distance at the deformed chisel.

Table 4.5
Comparison of Few Distances of the Chisel before and after Deformation

Parameters	Undeformed Shape	Deformed Shape
Delta X = $X_2 - X_1$	45.000 mm	45.002 mm
Delta Y = $Y_2 - Y_1$	−27.500 mm	−26.398 mm
Delta Z = $Z_2 - Z_1$	−200.000 mm	−200.034 mm
Distance L	206.840 mm	206.726 mm

normal and shear. When Hook's law is active, the normal stress causes normal strain, whereas shear stress causes shear strain. Normal strains produce dilatations, whereas shear strain produces angle deformations.

If we study a two-dimensional infinitesimal rectangular material element with dimensions $dx \times dy$, which after its deformation looks a parallelogram (Figure 4.33), we can write

$$length\ (AB) = dx \text{ and for the adjacent figure}$$

$$length\ (ab) = \sqrt{\left(dx + \frac{\partial u_x}{\partial x}dx\right)^2 + \left(\frac{\partial u_y}{\partial x}dx\right)^2} = dx\sqrt{1 + 2\frac{\partial u_x}{\partial x} + \left(\frac{\partial u_x}{\partial x}\right)^2 + \left(\frac{\partial u_y}{\partial x}\right)^2}.$$

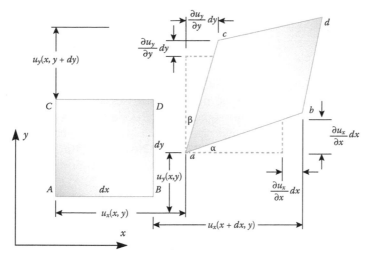

Figure 4.33
Geometric deformation of an infinitesimal material element.

For small displacement gradients, the expression above is simpler:

$$length\,(ab) \approx dx + \frac{\partial u_x}{\partial x}\,dx$$

Thus, the **normal strain** in the x-direction (**EPSX**) of the studied element is

$$\varepsilon_x = \frac{extension}{original\ length} = \frac{length\,(ab) - length\,(AB)}{length\,(AB)}$$

$$= \left(dx + \frac{\partial u_x}{\partial x}\,dx - dx \right) / dx = \frac{\partial u_x}{\partial x}\ (\text{EPSX})$$

And the normal strains in the y- and z-directions are $\varepsilon_y = \dfrac{\partial u_y}{\partial y}(\text{EPSY})$ and $\varepsilon_z = \dfrac{\partial u_z}{\partial z}(\text{EPSZ})$.

The **shear strain**, noted as γ_{xy}, defines the change in the square angle between AC and AB and is equal to $\gamma_{xy} = \alpha + \beta$. Further, based on the geometry, the following can be written:

$$\tan(\alpha) = \frac{\dfrac{\partial u_y}{\partial x}\,dx}{dx + \dfrac{\partial u_x}{\partial x}\,dx} = \frac{\dfrac{\partial u_y}{\partial x}}{1 + \dfrac{\partial u_x}{\partial x}} \text{ and } \tan(\beta) = \frac{\dfrac{\partial u_x}{\partial y}\,dy}{dy + \dfrac{\partial u_y}{\partial y}\,dy} = \frac{\dfrac{\partial u_x}{\partial y}}{1 + \dfrac{\partial u_y}{\partial y}}.$$

For small displacement gradients, we have $\dfrac{\partial u_x}{\partial x} \ll 1$ and $\dfrac{\partial u_y}{\partial y} \ll 1$, and for small rotations, $\tan(\alpha) \approx \alpha$ and $\tan(\beta) \approx \beta$.

Consequently: $\alpha \approx \dfrac{\partial u_y}{\partial x}$, $\beta \approx \dfrac{\partial u_x}{\partial y}$ and $\gamma_{xy} = \alpha + \beta = \dfrac{\partial u_y}{\partial x} + \dfrac{\partial u_x}{\partial y} = \gamma_{yx}$ (**GMXY**).

Additionally: $\gamma_{yz} = \dfrac{\partial u_y}{\partial z} + \dfrac{\partial u_z}{\partial y} = \gamma_{zy}$ (**GMYZ**) and $\gamma_{xz} = \dfrac{\partial u_x}{\partial z} + \dfrac{\partial u_z}{\partial x} = \gamma_{zx}$ (**GMXZ**).

If we rotate the infinitesimal element around its geometric centre, it is possible to orientate it in a way that there will be no shear stresses along the sides AB and AC. Thus, the square angles will remain square after the deformation, only two perpendicular dilatations, due to the normal stresses that will be observed. These normal stresses are known as principal stresses (**P1**(σ_1), **P2**(σ_2) and **P3**(σ_3)) and are related to the minimum and maximum stretches. The corresponding strains are known as **principal strains** (**E1**(ε_1), **E2**(ε_2) and **E3**(ε_3)).

Equivalent stress (**VON**: *von Mises stress*) is often used in the design work because it allows any arbitrary three-dimensional stress state to be represented by a single positive stress value. Equivalent stress is part of the maximum equivalent stress failure theory used to predict yielding in a ductile material. Equivalent stress is related to the principal stresses by the equation $\sigma_e = \sqrt{\dfrac{(\sigma_1 - \sigma_2)^2 + (\sigma_2 - \sigma_3)^2 + (\sigma_3 - \sigma_1)^2}{2}}$.

The von Mises or **equivalent strain** ε_e (**ESTRN**: Equivalent strain) is computed as

$\varepsilon_e = \dfrac{1}{1+\nu} \sqrt{\dfrac{(\varepsilon_1 - \varepsilon_2)^2 + (\varepsilon_2 - \varepsilon_3)^2 + (\varepsilon_3 - \varepsilon_1)^2}{2}}$, where ν is Poisson's ratio.

A three-dimensional linear elastic solid with loads supplied by external forces F_i and through support reactions can be considered to be made up of small cubic elements as shown in Figure 4.34.

The incremental strain energy **dU** for this elemental cube of volume **dV** can be written as

$$dU = \frac{1}{2}\{\sigma_x \varepsilon_x + \sigma_y \varepsilon_y + \sigma_z \varepsilon_z + \sigma_x \varepsilon_x + \tau_{xy} \gamma_{xy} + \tau_{yz} \gamma_{yz} + \tau_{zx} \gamma_{zx}\}dV$$

Integrating the incremental strain energy **dU** over the entire volume **V**, the **total strain energy**, **U** (**ENERGY**: Total strain energy), is obtained:

$$U = \frac{1}{2}\int_V \{\sigma_x \varepsilon_x + \sigma_y \varepsilon_y + \sigma_z \varepsilon_z + \sigma_x \varepsilon_x + \tau_{xy} \gamma_{xy} + \tau_{yz} \gamma_{yz} + \tau_{zx} \gamma_{zx}\}\, dV.$$

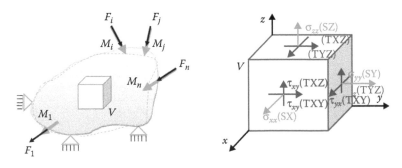

Figure 4.34
Stresses within a linear elastic solid.

Therefore, the energy stored in a body due to deformation is called **strain energy**. **Strain energy density** (**SEDENS**: Strain energy density) is a measure of how much energy is stored in small volume elements throughout a material. In other words, it is a scalar function equal to the strain energy per unit volume (Figure 4.35).

All these plots are defined by default through

Simulation → Options → Default Options → Plot → Default Plots → Static Study Results

Of course, as has been discussed before, a few more plots can be defined using that path and this will affect all analysis. But it is much easier to open some of the plots suggested by the **Results** panel (Figure 4.14b) of the **SW Simulation analysis tree**. It can be displayed by right clicking on the **Results** folder (Figure 4.14a).

The first command to be discussed in detail is the **Define Factor of Safety Plot** (🔣). It has already been briefly discussed, and some plots were given (Figure 4.10). The **Factor of Safety** property manager opens after clicking on the command line. It consists of three steps, and you can navigate among the windows by clicking the arrows at the top of the property manager (🔵 and 🔵, 1, Figure 4.36a).

The first window combines two sub-windows. The first sub-window is titled **Step 1 of 3** (2, Figure 4.36a) and the second is **Property** (3, Figure 4.36a). While similar **Property** sub-window has already been discussed (Figure 4.16e and f), the **Step 1 of 3** sub-window is unknown to us. It guides the user through defining the factor of safety criteria and creating the plot. At first, we must choose whether the factor of safety

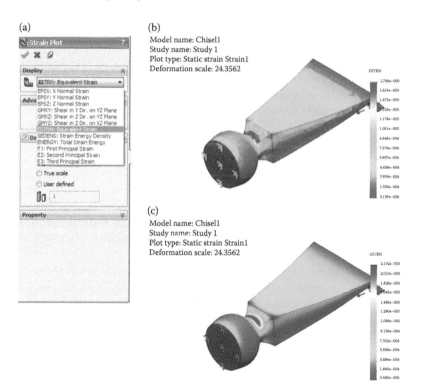

Figure 4.35
Equivalent strain plot of the deformed chisel. (a) Display window; (b) equivalent strain – element mode; (c) equivalent strain – node mode.

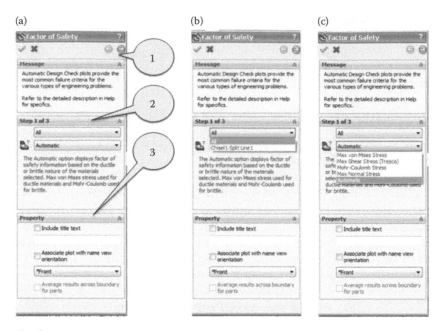

Figure 4.36
Factor of Safety window – Step 1 of 3. (a) Step 1 of 3 window; (b) defining the scope of the safety calculations; (c) defining the method of the safety calculations.

calculations will include all components of the model (**All**) or only a particular one (🐾, Figure 4.36b). The program considers the stress limit only of selected components; all other components are coloured in grey in the plots. The program cannot calculate the factor of safety of those components whose material has no defined stress limit.

The next step is to select the failure criterion (📘, Figure 4.36c). The failure criterion predicts the failure of a material subjected to multi-axial state of stresses. Regarding the selected criterion, the safety reserves of the model can be assessed. Materials behave either in a ductile or in a brittle manner depending on the temperature, loading, etc. Until now, a failure criterion applicable to all materials in all cases and conditions has not been developed. That is why you can choose and try a set of criteria that suit you best. Possible choices for solid bodies are **Max von Mises Stress**, **Max Shear Stress (Tresca)**, **Mohr–Coulomb Stress**, **Max Normal Stress** or **Automatic**. Three more criteria, available for composite shells only, **Tsai-Hill Criterion**, **Tsai-Wu Criterion** and **Max Stress Criterion**, will not be discussed now.

The **Max von Mises Stress** criterion is suitable **for ductile materials**. It is based on shear-energy theory, which states that a ductile material starts to yield at a location, where and when the elastic energy of distortion reaches a critical value. At that moment, von Mises stress $\left(\sigma_{von\ Mises} = \sqrt{\dfrac{(\sigma_1 - \sigma_2)^2 + (\sigma_2 - \sigma_3)^2 + (\sigma_3 - \sigma_1)^2}{2}} \right)$ becomes equal to the stress limit. In most cases, the yield strength is used as the stress limit. However, the software allows the user to use the ultimate tensile or another stress limit and to compare $\sigma_{von\ Mises}$ to σ_{limit}. The factor of safety at a location is calculated as Factor of Safety (FoS) $= \dfrac{\sigma_{limit}}{\sigma_{von\ Mises}}$. In the case of pure shear, von Mises stress can be expressed as $\sigma_{von\ Mises} = \sqrt{3}\tau$. Hence, failure occurs when $\tau_{max} = 0.577\ \sigma_{yield}$. As the yield strength is a temperature-dependent property, the temperature of the component is also indirectly considered.

The **Max Shear Stress** criterion, also known as the **Tresca** criterion, is suitable **for ductile materials**. It is based on the maximum shear stress theory, which specifies that a material would flow plastically if $\sigma_{Tresca} > \sigma_{limit}$. Denoting the greatest of the three principal shear stresses τ_{max} and knowing that $\tau_{max} = \tau_{13} = \dfrac{\sigma_1 - \sigma_3}{2}$ and $\sigma_{Tresca} = \sigma_1 - \sigma_3 = 2\tau_{max}$, the Tresca failure criterion can be written as $\tau_{max} \geq \dfrac{\sigma_{limit}}{2}$. Hence, the Tresca factor of safety is Factor of Safety (FoS) $= \dfrac{\sigma_{limit}}{2 * \tau_{max}}$.

The next possible choice is the **Mohr–Coulomb Stress** criterion, which is based on the internal friction theory. This criterion is used **for brittle materials** with different tensile and compressive properties. As brittle materials do not have a specific yield point, the yield strength cannot be used to define the limit stress. Usually the values of limit stress are based on the **Tensile Strength** and **Compressive Strength** of the material. For most of the brittle materials, $-\sigma_{Compressive\_Strength} \gg \sigma_{Tensile\_Strength}$. This theory predicts failure to occur if $\sigma_3 > 0$, that is, all normal stresses are tensile, when $\sigma_1 \geq \sigma_{Tensile\_limit}$; if $\sigma_1 < 0$, that is, all normal stresses are compressive, when $\sigma_3 \leq -\sigma_{Compressive\_limit}$; and if $\sigma_1 \geq 0$ and $\sigma_3 \leq 0$, when $\dfrac{\sigma_1}{\sigma_{Tensile\_limit}} + \dfrac{\sigma_3}{-\sigma_{Compressive\_limit}} \geq 1$. Thus, the Mohr–Coulomb factor of safety is

$$\text{Factor of Safety (FoS)} = \frac{1}{\dfrac{\sigma_1}{\sigma_{Tensile\_limit}} + \dfrac{\sigma_3}{-\sigma_{Compressive\_limit}}}.$$

The **Max Normal Stress** criterion is based on the maximum normal stress theory. The criterion is **used for brittle materials**. According to the **Maximal normal stress** theory, failure occurs when the maximum principal stress σ_1 reaches the ultimate strength of the material for simple tension, that is, this theory predicts failure to occur when $\sigma_1 \geq \sigma_{limit}$. Hence, Factor of Safety (FoS) $= \dfrac{\sigma_{limit}}{\sigma_1}$.

Before continuing, it is better to systematise and compare these four failure criteria (Table 4.6).

There is one more choice in the set of failure criteria (📖, Figure 4.36c) and this is the most commonly used **Automatic** choice. Then the program selects the most appropriate criterion in relation to the **Default Failure Criterion** assigned in the **Material** dialog box for each material (Figure 2.30). If no criterion is assigned, the program starts the **Mohr–Coulomb Stress** criterion. For the **Max von Mises Stress** or **Max Shear Stress (Tresca)** criterion, the program uses **yield strength** as allowable stress, while for the **Max Normal Stress** or for **Mohr–Coulomb Stress** criterion, the program uses **tensile strength** as allowable stress.

After selecting the failure criterion, we move to the next window by clicking on the right navigation arrow (1, Figure 4.36a). Thus, **Step 2 of 3** is activated (Figure 4.38). According to the preselected Step 1 criterion, one of the three windows shown in Figure 4.38 opens. The window in Figure 4.38a is active for **Max von Mises Stress**, **Max Shear Stress (Tresca)** and **Max Normal Stress** criteria. The second window (Figure 4.38b)

Table 4.6
Comparison of the Built-In Failure Criteria

Failure Criteria	Factor of Safety (FoS)	Materials and Assumptions	
Max von Mises Stress	$FoS = \dfrac{\sigma_{limit}}{\sigma_{von\ Mises}}$ $\sigma_{von\ Mises} =$ $\sqrt{\dfrac{(\sigma_1 - \sigma_2)^2 + (\sigma_2 - \sigma_3)^2 + (\sigma_3 - \sigma_1)^2}{2}}$	Used for ductile materials	It is experimentally proved that for pure shear, the Max von Mises stress criterion provides more accurate results than the Tresca criterion
Max Shear Stress (Tresca)	$FoS = \dfrac{\sigma_{limit}}{2 * \tau_{max}}$ $\tau_{max} = \dfrac{\sigma_1 - \sigma_3}{2}$		As a model considering the plastic material behaviour, Tresca's criterion is more conservative than the Max von Mises criterion (Figure 4.37)
Mohr–Coulomb Stress	$FoS = \dfrac{1}{\dfrac{\sigma_1}{\sigma_{Tensile\_limit}} + \dfrac{\sigma_3}{-\sigma_{Compressive\_limit}}}$	Used for brittle materials	It assumes different ultimate strengths of the material in tension and compression
Max Normal Stress	$FoS = \dfrac{\sigma_{limit}}{\sigma_1}$		It assumes that the ultimate strength of the material in tension and compression is the same, which is not valid in a lot of cases

is assigned to the **Mohr–Coulomb Stress** criterion, and the last window (Figure 4.38c) opens after selecting the **Automatic** option. At first, the **Units** of the stress limit must be set (1a, Figure 4.38a; or 2a, Figure 4.38b), and after that the **Stress** limit itself. We can select among the following options: **Yield strength** (recommended for ductile materials); **Ultimate strength** (recommended for brittle materials); or **User defined** (1b,

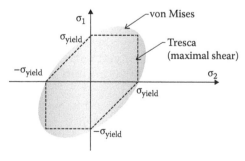

Figure 4.37
Projection of the von Mises and Tresca failure criteria into a plane.

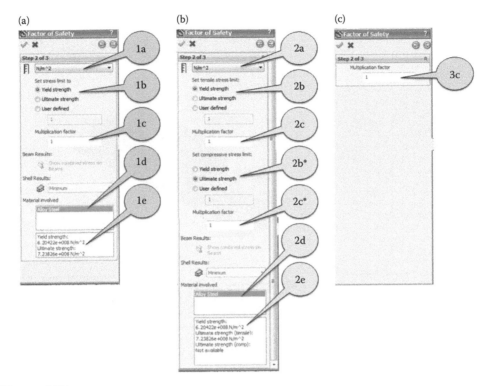

Figure 4.38
Factor of Safety panel – Step 2 of 3. (a) Selecting stress units; (b) selecting stress limit; (c) selecting Multiplication factor.

Figure 4.38a; or 2b, Figure 4.38b). The **Multiplication factor** is used to multiply the stress limit in order to increase or to lower the limit. Its default value is 1.0 (1c, Figure 4.38a; 2c, Figure 4.38b; or 3c, Figure 4.38c). For the **Mohr–Coulomb Stress** criterion, two different sets for compressive and for tensile stresses are introduced (2b and 2c; 2b* and 2c*, Figure 4.38b). The **Material involved** box shows the material of the component (1d, Figure 4.38a and 2d, Figure 4.38b) and its properties related to the stress limits used by the program to calculate the **Factor of Safety** (1e, Figure 4.38a and 2e, Figure 4.38b).

The last window is **Step 3 of 3** (Figure 4.39a). We can select the **Factor of safety distribution** to plot the distribution of factor of safety (Figure 4.10) or **Areas below factor of safety** and to input a value for the factor of safety to be 5, for example (Figure 4.39a). The **Safety result** box shows the minimum factor of safety based on the selected criterion. The **Factor of safety (FoS)** property manager closes by clicking **OK** (✓), and the plot is displayed in the **Graphics area** (Figure 4.39b). The program displays all areas with FoS less than the specified value of 5 in **red** (unsafe regions) and regions with higher FoS in **blue** (safe regions).

The analysed chisel is made of steel, which is a ductile material. Nevertheless, if the **Max von Mises Stress** criterion is assigned to the applied material (Figure 2.30), we can vary the criteria and compare the results. Both the **Max von Mises Stress** and **Max Shear Stress** criteria are used for ductile materials. Additionally, the **Automatic** option is added to the set of compared results. All compared results are given in Table 4.7. They confirm the previous statement that the **Max Shear Stress** criterion is more conservative than the **Max von Mises Stress** criterion (Table 4.6).

(a) (b)

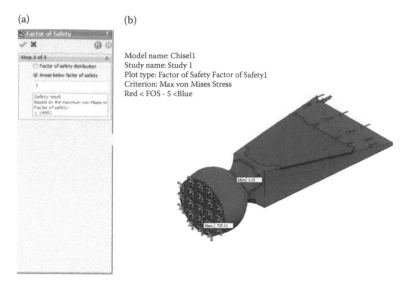

Model name: Chisel1
Study name: Study 1
Plot type: Factor of Safety Factor of Safety1
Criterion: Max von Mises Stress
Red < FOS - 5 <Blue

Figure 4.39
Factor of Safety panel – Step 3 of 3. (a) Step 3 of 3 window; (b) areas with factor of safety below 5 according to Max von Mises Stress criterion.

Table 4.7
Comparison of Minimal FoS Calculated According to Different Criteria

Failure Criterion	Min FoS	Max FoS
Max von Mises Stress	$FoS_{min} = 1.15$	$FoS_{max} = 707.11$
Max Shear Stress	$FoS_{min} = 1.06$	$FoS_{max} = 627.91$
Automatic (uses Max von Mises Stress in this case)	$FoS_{min} = 1.15$	$FoS_{max} = 707.11$

Two more plots can be defined through the **Results** panel (Figure 4.6d) of the **SW Simulation analysis tree**. These are the **Define Design Insight Plot** (Figure 4.11) and the **Define Fatigue Check Plot**. As we have already gotten accustomed to them while discussing the different steps of viewing the results through the **Results advisor**, they will not be discussed here again.

Viewing the results through the **Results** pop-up menu (Figure 4.6c) of the **SW Simulation analysis tree** enlarges extensively the options to systematise them compared to the **Results Advisor** guidance (Figure 4.6d).

We have defined three basic plots through

Simulation → Options → Default Options → Plot → Default Plots

They are **Stress**, **Displacement** and **Strain**.

Their visual presentation is discussed in detail, following the plot options in Figure 4.15 property menus. They include

- **Edit Definition** (Figures 4.16 and 4.18)
- **Animate** (Figure 4.19)

- **Section Clipping** (Figures 4.20 and 4.21)
- **Iso Clipping** (Figures 4.22 and 4.23)
- **Chart Options** (Figures 4.24 and 4.25)
- **Settings** (Figures 4.26 and 4.27)
- **Probe** (Figure 4.28)
- **Deformed Result**
- **Standard View toolbar** (Figures 4.29 and 4.30)

The operation of all these commands is explained in the context of the **Stress** plot. Some additional explanations, particularly for **Displacement** (Figures 4.31 and 4.32) and **Strain** plots (Figure 4.35), are added.

Finally, the procedure of **Factor of Safety (FoS)** calculations and corresponding plots (Figures 4.36, 4.38, and 4.39) are discussed.

During that stage, we learned how to

- View results through the Results pop-up menu of the SW Simulation analysis tree
- Change the visual and data properties of the plots
- Develop Factor of Safety (FoS) calculations and generate corresponding plots

We learned some basic theory about

- Strain/Displacement/Stress relations and different types of strain (normal, shear, principal, equivalent strain energy and strain energy density)
- Different methods for FoS calculations and compared them

4.2.3   Results Display through Icons on the SW Simulation Command Bar

The last manner of plotting the results is by using the icons on the **SW Simulation command bar** (Figure 4.6b). As we have already gotten accustomed to most of the icons displayed at the command bar, their detailed presentations will be omitted.

By clicking the arrow below the **Results Advisor** icon on the command bar (⬛, 1, Figure 4.40), a pop-up menu appears. The first icon in that menu directly activated the **Results Advisor** bar (⬛, 2a, Figure 4.40) on the right side of the **Graphics area** to guide us (see Section 4.2.1). Unknown to us is the third command line **New Plot**

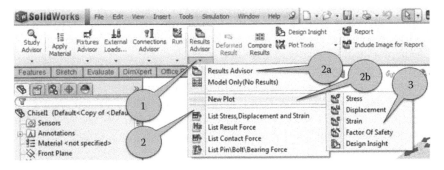

Figure 4.40
Results Advisor on SW Simulation command bar.

(2b, Figure 4.40). It starts a new menu (3, Figure 4.40) that includes all previously discussed commands such as **Stress** (⬛), **Displacement** (⬛), **Strain** (⬛), **Factor of Safety** (⬛) and **Design Insight** (⬛). Selecting any of them, the corresponding property manager opens and the program waits for the input of the plot options. The operation of these property managers has been explained in Chapter 3.

Clicking the **Deformed Result** (⬛) controls the overlapping of the plot on the deformed or undeformed shape of the model.

The **Compare Results** command (⬛, Figure 4.40), which is yet to be discussed, allows comparing up to four results in each of the following ways:

- **Comparing the current plot to similar plots from other studies**: This supposes few static studies to be done. There is no limit in the number of the developed studies, but no more than four could be compared simultaneously. We have chosen to develop a second study with different loads. A uniform **Pressure** load of 100 MPa is applied at the head face of the chisel. No loads at the lateral sides of the chisel are considered. After running the analysis, the two studies are ready to be compared.

  To do this, all steps given below should be completed:
  - Show the compared plot in one of the studies (Figure 4.41a).
  - Click the **Compare Results** icon (⬛, Figure 4.40). The **Compare Results** property manager opens.
  - Select **Compare selected result across studies** in the **Options** sub-window (Figure 4.41b).
  - Select compared studies, no more than four. If there are four or fewer studies with similar plots, they are all selected by default. In our case, **Study 1** and **Study 2** are selected by default (Figure 4.41b).
  - Click **OK** (✓).
  - The plots appear in different panes – two horizontal panes in our case (Figure 4.41d). Clicking on a pane displays the associated **Simulation study tree** – **Study 1** for the first pane and **Study 2** for the second. If there are no similar plots in all studies, the program creates temporary plots and removes them after exiting the comparison.
  - Click the **Exit Compare** button on the small newly opened **Compare results** window (Figure 4.41c) to close the comparing plots.
- **Comparing multiple plots from the current study**

  To do this, the steps given below should be completed:
  - Click the **Compare Results** icon (⬛, Figure 4.40). The **Compare Results** property manager opens.
  - Select the **View multiple results of current study** in the **Options** sub-window (Figure 4.42a). It selects the compared plots, yet no more than four plots can be selected. The first four plots are selected by default. The order of the display by the **Compare Results** property manager plots corresponds to the order in the **SW Simulation analysis tree** (Figure 4.42a).
  - Select the **Use settings from this plot for plots of the same type** (Figure 4.42a) and select a plot in the **Study name** to use this plot as a sample and to synchronise temporarily the settings of all plots. This option is not selected for the shown comparison (Figure 4.42b).
  - Click **OK** (✓).

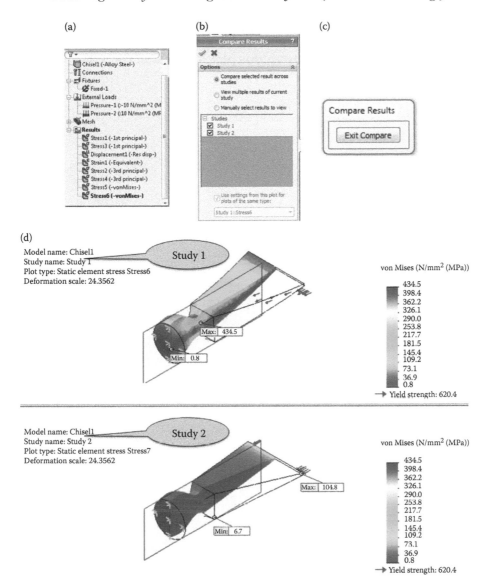

Figure 4.41
Comparison of similar plots of different studies. (a) Active Stress plot; (b) Exit Compare window; (c) compared similar plots; (d) cut plots of von Mises stresses.

- The plots appear in different panes – four panes for four selected plots (Figure 4.42b).
- Click **Exit Compare** (Figure 4.41c) to exit the comparison.
- **Comparing arbitrary plots from a different study**
  The necessary steps are given below:
  - Click the **Compare Results** icon ([⊞], Figure 4.40).
  - Select **Manually select results to view** in the **Options** sub-window (Figure 4.43a).

- Select up to four compared plots from different studies and type. We select **Displacement** and **Strain** plots of **Study 1** and **Study 2** (Figure 4.43a).
- Optionally, **Use settings from this plot for plots of the same type** can be selected (Figure 4.43a).
- Click **OK** (✅).

(a)   (b)

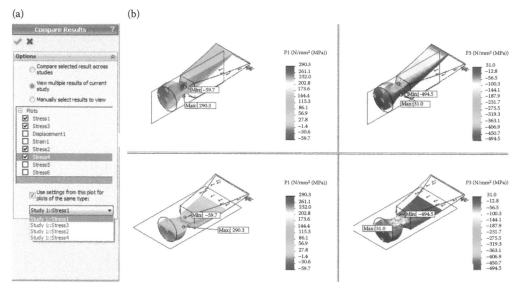

Figure 4.42
Viewing multiple plots from the current study. (a) Selection of plots; (b) cut plots of maximum and minimum principal stresses for Study 1.

(a)   (b)

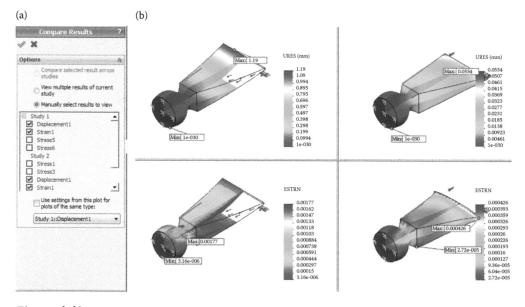

Figure 4.43
Comparison of arbitrary plots from different studies. (a) Selection of plots; (b) angle cut plots of displacement and equivalent strain for Study 1 (in the left) and for Study 2 (in the right).

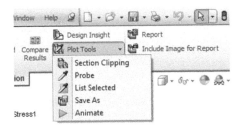

Figure 4.44
Plot Tools pop-up menu.

- The plots appear in different panes (Figure 4.43b).
- Click **Exit Compare** (Figure 4.41c).

If a plot is active, clicking **Plot Tools** (▨, Figure 4.44) opens a pop-up menu that combines some of the previously discussed commands for defining plot properties. They are **Section Clipping** (▥), **Probe** (▨), **Animate** (▷) as well as **List Selected** (▨) and **Save As** (▨).

Clicking on the **Report** icon (▨, Figure 4.44) opens the known **Report Options** panel (Figure 4.4b).

Include Image for Report (▨, Figure 4.44) inserts the image into the report through the opened **Insert Image** property manager (Figure 4.45).

- The **Image Type** sub-window (Figure 4.45a) enables us to choose between two options. Our choice can be either the **Current View**, which inserts the current plot of the model, mesh or results, viewed in the **Graphics area**, or the **Image file**, which inserts an already existing picture of *.jpg, *.bmp or *.gif

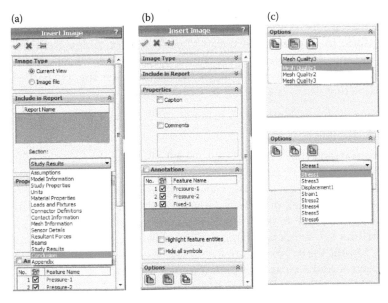

Figure 4.45
Insert Image property manager. (a) Image Type and Include in Report sub-windows; (b) Properties and Annotations sub-windows; (c) Options sub-window.

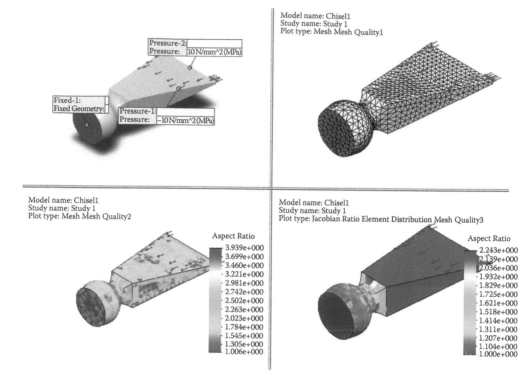

Figure 4.46
Different views of the meshed chisel.

types saved anywhere in our computer. If the state of the model changes, during the analysis, the embedded pictures do not update automatically.

- The **Include in Report** sub-window (Figure 4.45a) enables us to choose the report and the exact section where the image is to be inserted.
- The **Properties** sub-window (Figure 4.45b) handles the properties of the inserted picture, such as **Caption** or **Comments**.

The next two sub-windows are accessible only if the **Current View** option is selected at the beginning.

- The **Annotations** sub-window (Figure 4.45b) handles the viewing of the loads (Pressure – 1 and Pressure – 2), fixtures (Fixed – 1) or connectors (see the plot at the top left of Figure 4.46). The viewed information can include the type and value (**Feature Name**), the effected entities (**Highlight feature entities**) and the symbols that denote the feature (**Hide all symbols**).
- The **Options** sub-window (Figure 4.45c) helps to select the type of the plot.
  - The **Model View** (⬛, Figure 4.45c) captures the current model view and displays it in the **Graphics** area (the plot at the top left, Figure 4.46).
  - The **Mesh View** (⬛, Figure 4.45c) option is available only if the model is meshed. There are three views of the mesh (Figure 4.46): **Mesh Quality 1** – displays the meshed model (the plot at the right top); **Mesh Quality 2** – displays the Aspect ratio (the plot at the left bottom);

Mesh Quality 3 – displays the Jacobian Ratio Element Distribution (the plot at the right bottom).
- **Result Plots** (▣, Figure 4.45c) can be activated only if there are any results. It allows selection of a generated result plot, which is included in the selection list.
- To see the plots in the **Graphics area**, the **Activate** button must be clicked.
- The **Insert Image** property manager closes by clicking **OK** (✔).

In this section, we used the icons on the **SW Simulation command bar** (Figures 4.6b and 4.40) to systematise and display more results.

We discussed the three different ways of visually comparing the results using the **Compare Results** command:

- **Comparison of similar plots of different studies** (Figure 4.41)
- **Viewing multiple plots from the current study** (Figure 4.42)
- **Comparison of arbitrary plots from different studies** (Figure 4.43)

We deepen our knowledge of how to insert an image in a report using the **Insert Image** property manager (Figure 4.45). The options of that command allow the insertion of either a **Current View** (Figure 4.45a) from the **Graphics area** or an existing **Image file** (Figure 4.45a) by browsing the directories of the computer.

During that stage, we used the icons on the SW Simulation command bar to display some more plots of viewing the results.

We learned how to

- Compare up to four plots of one and the same or of different studies (▦)
- Insert a current plot or an already saved image file in the report (▣)
- Superimpose the plot either over the undeformed or over the deformed shape of the body in an easier manner (▨)

4.3 LISTING THE RESULTS OF THE ANALYSIS

While plotting the results provides better view of the component distribution, their listing provides a higher precision. The program presents the lists in tables. The main weakness of the listed results is their quantity. Therefore, sometimes it is very difficult for these data to be sorted and analysed.

Without any concerns about the type of the analysis, we can **list mesh properties** using the **Probe** property manager (▨) and clicking the **Save** icon (▣) on the **Report Options** sub-window. Based on selecting the **Advanced Option** sub-window mode, element or node, the program saves a table including information on the number of elements/nodes and its global coordinates.

With regards to focusing on the static studies of a single body, there are a few commands with regards to listing the results: to list stress values, to list displacements, to list strain and to list the resultant force. All these commands can be started either from the pop-up **Results** menu from the **SW Simulation** analysis tree (Figure 4.47a) or from the pop-up **Results Advisor** menu from the **SW Simulation command bar** (Figure 4.47b).

All files generated by the program lists are either **MS Excel files** (*.csv) or **text files** (*.txt). The first are not typical MS Excel files, and it is not possible to directly filter the data, generate charts, etc.; but they are easy to be imported in the SW environment. The text files can be opened easily with any text editor, as well as with the MS Excel software; thus, they behave as typical MS Excel files and their data are easy to be sorted, filtered, etc.

The command to be discussed here is the **List Stress, Displacement and Strain** command (, Figure 4.47b). It opens the **List Results** property manager (Figure 4.48a), which includes four sub-windows: **Quantity**, **Component**, **Advanced Options** and **List Set**.

- The first to be explained is **how to list stresses**. We have to follow the stages stated in the following:
  - Check **Stress** in the **Quantity** sub-window (Figure 4.48b).
  - Select the listed component (, Figure 4.48b) from the provided list. If you choose a directional stress, such as normal or shear stress, you can pick an alternative reference entity in the **Advanced Options** sub-window. Otherwise, this window is not accessible. Next is to set the **Units** ( ) of the list. **Plot Step** ( ) sets the step number at which the selected results are to be listed. Thus, it is not active for static studies.
  - The first optional stage in the **Advanced Options** sub-window (Figure 4.48d) is picking a reference entity ( ), if appropriate. As has been explained, it can be a plane, an axis or a newly defined coordinate system. If no reference entity is picked, the program uses the global coordinate system to orientate the stresses. By selecting **Nodes**, the software is

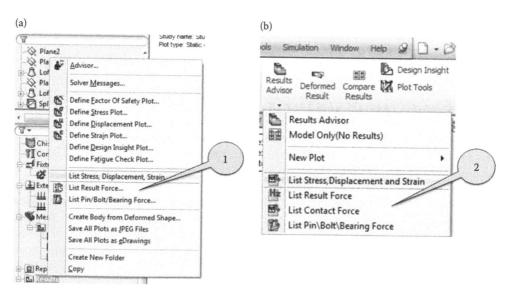

Figure 4.47
Different ways to start a list command. (a) Pop-up Results menu; (b) pop-up Results Advisor menu.

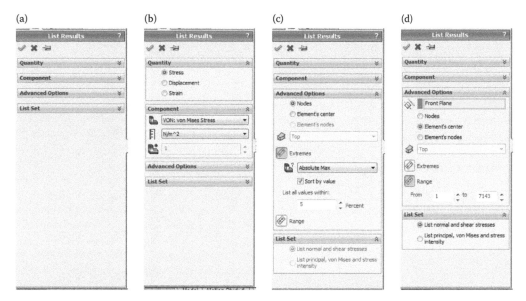

Figure 4.48
List Results property manager – stresses. (a) List Results property manager – basic view; (b) Quantity and Component sub-windows for creation of stress list; (c) Advanced Options sub-window for extreme stress list; (d) Advanced Options and List Set sub-windows for range stress list.

being informed to list the selected stress in the nodes, while selecting the **Element's center** lists the calculated stresses for the element mode. This means that each element value is an averaged value of the stresses fixed to the element nodes. The **Element's nodes** option is available only if **Range** is selected.
- The next **Advanced Options** choice is **Extremes** (⬛) or **Range** (⬛) values to be included in the list. Checking **Extremes** (⬛, Figure 4.48c) searches for extreme values, and if **Sort by value** is selected, it sorts the list according to the following criteria (⬛): **Absolute Max**, **Algebraic Max** and **Algebraic Min**. **List all values within** shortens the list by including only these values, which are in the range, defined by 5% (by default) tolerance of the extreme value. **Range** (⬛, Figure 4.48d) lists the values within a range, specified by the lowest and the highest node/element numbers. The type of the listed data in the **Range** mode is introduced by checking the **List Set** sub-window (Figure 4.48d). The program generates either a **List of normal and shear stresses** or lists of principal stresses, von Mises stresses and stress intensity.
- The adjusted list is generated after clicking **OK** (✔).
- The **list of displacements** can be generated as follows:
  - Our first choices are the **component** (⬛), from the displacement list, and the **units** (⬛) (Figure 4.49a). The rules that must be observed are similar to those for the generation of stress list.
  - **Advanced Options** (Figure 4.49a), particularly **reference entity** (⬛), **Extremes** (⬛) and its adjusting options, obey the same rules as the

(a)　　　　　　　　　　(b)　　　　　　　　　(c)　　　　　　　　　(d)

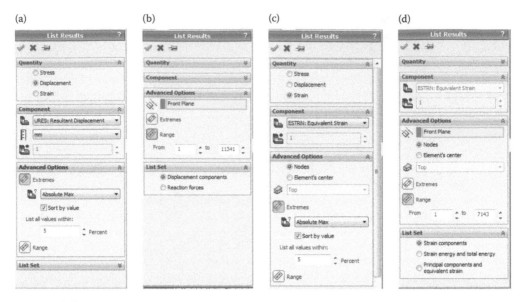

Figure 4.49
List Results property manager – displacements. (a) Quantity, Components and Advanced Options sub-windows for extreme displacement list; (b) Advanced Options and List Set sub-windows for range displacement list; (c) Quantity, Components and Advanced Options sub-windows for extreme strain list; (d) Advanced Options and List Set sub-windows for range strain list.

corresponding options for stresses. As only the displacements of the nodes are calculated, there is no element mode. Checking the **Range** (⬦) sets the range of the list based on the numbers of the nodes. It relates the **List Set** (Figure 4.49b) options **Displacement components** or **Reaction forces** lists.

- The displacement list will be generated after clicking **OK** (✔), and the corresponding tables for **displacement components** (UX, UY, UZ, UREZ) and for **reaction force components** (RFX, RFY, RFZ, RFREZ) are shown in Figure 4.50.
- Compared to the lists of stresses and of displacements, the generation of **list of strain** differs in the following:
  - The list of the **components** (🔖) and the **units** (☰) (Figure 4.49c).
  - **Nodes** and the **Element's center** handled through **Advanced Options** (Figure 4.49c and d) are available in **Extremes** (⬦) and in **Range** (⬦) lists.
  - Generation of **Range** (⬦) lists is again directly related to the options in the **List Set** sub-window (Figure 4.49d). These options are as follows: **Strain components**, which if selected lists the normal and the shear strain components with respect to the selected reference geometry, if any; and **Strain energy and total energy**, which lists strain energy density and total strain energy and **Principal components and equivalent strain**.
  - The **List Results** property manager closes and the list command is executed after clicking **OK** (✔).

(a) 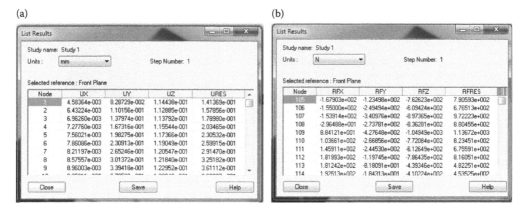 (b)

Figure 4.50
Displacement range tables. (a) Displacement components table (mm); (b) reaction force components (N).

The next command to be explained is the **List Result Force** (⬛, Figure 4.47a or ⬛, Figure 4.47b). Clicking its icon opens the **Result Force** property manager (Figure 4.51a), which combines five sub-windows:

- The **Options** sub-window (Figure 4.51b) limits our choice to **Reaction force**, **Free body force** and, inaccessible for that study, **Remote load interface force Contact/Friction force**. Selecting the **Reaction Force** lists X, Y, Z components and the magnitude of the force for selected entities or for the entire model. Checking the **Free body force** lists free body forces due to the action of external loads, restraints, contacts, etc. at all preselected entities, such as faces, edges, etc.

(a)   (b)   (c)   (d)   (e)

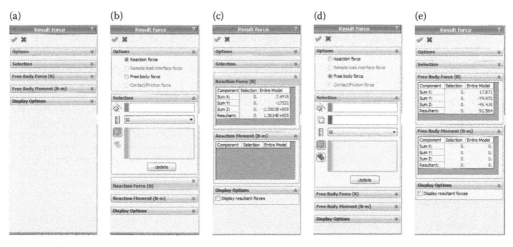

Figure 4.51
Result Force property manager. (a) Result Force window; (b) Options and Selection sub-windows – Reaction force selected; (c) Reaction Force, Reaction Moment and Display Options sub-windows; (d) Options and Selection sub-windows – Free body force selected; (e) Free Body Force, Free Body Moment and Display Options sub-windows.

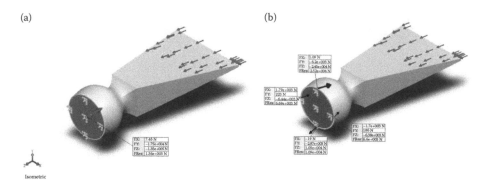

Figure 4.52
Free body force diagrams. (a) Selected root face entity; (b) checked Display Options.

- As the options in the **Selection** sub-window related to the generation of **Reaction force** (Figure 4.51b) and **Free body force** (Figure 4.51d) lists are almost similar, they will be discussed simultaneously. We can define a new **reference entity** – plane, axis or coordinate system ( ) or a **vertex or reference point for location of the moment** ( ). After that, we select the output **units** ( ). The list of picked **entities** is displayed in the blue window. They can be faces, edges or vertices ( ) or components ( ) for which the forces are to be listed. Finally, the **Update** button must be clicked to update newly selected properties. Generally, the components for any selected entity are displayed, and their values and the value of the magnitudes are directly written besides the model (Figure 4.52a).
- **Reaction Force (N)** and **Free Body Force (N)** (Figure 4.51c and e) display the components (**Sum X, Sum Y and Sum Z**) and the magnitudes (**Resultant**) of the reaction moment/free body force of the selected entities or of the entire part (Figure 4.52).
- **Reaction Moment (N-m)** and **Free Body Moment (N-m)** (Figure 4.51c and e) display the components and the magnitude of the reaction moment/free body moment of the selected entities or of the entire model.
- When selected, **Display Options** (Figure 4.51c and e) displays the resultant force/free body force vector as a black arrow (Figure 4.52b).
- At the end, click **OK** ( ) to close the property manager.

In this section, we generated different lists of results. All useful icons can be reached in two ways:

- **Through the SW Simulation analysis tree:**

  Results (right click) → List Stress, Displacement, Strain/List Result Force...

- **Through the SW Simulation command bar:**

  Results Advisor (click the arrow below) → List Stress, Displacement, Strain/List Result Force...

We can save the lists in two different interchangeable file formats – *.csv and *.txt. Each format supposes different ways to filter and systematise the data in tables and diagrams that are suitable for future analysis (Figures 4.50 and 4.52).

We explained in details the operation of

- **List Results property manager** (Figure 4.49)
- **Result Force property manager** (Figure 4.51)

We learned how to generate and save different lists of results. All the lists can easily be opened and operated by different software packages, including commonly used programs of the MS Office package and text editors Wordpad and Notepad. We learned how to filter and arrange automatically the useful data.

We learned how to create and view

- Lists of stresses
- Lists of displacements
- Lists of strain
- Lists and diagrams of reaction forces and moments
- Lists and diagrams of free body forces and moments

4.4 DRAWING GRAPHS OF THE ANALYSIS RESULTS

There are some different graphs that the program generates while running static analysis. They can be classified in two major groups:

- **Convergence graphs** – these graphs present the convergence of the calculation process. They were partially discussed in Chapter 3 (Figure 3.16a). They are the program's visual feedback on the speed and convergence of the calculations. For example, to fasten the calculations, we can either set a higher threshold or decrease the number of iterations. If an adaptive method is selected and calculations are successful, a new command line is accessible in the **Results** pop-up menu – **Define Adaptive Convergence Graph** (Figure 4.53a). Clicking on it opens the **Convergence Graph** property manager, which includes only one sub-window and five checks.

    Optional selections for the **p-adaptive method** are as follows (Figure 4.53b):
    - Maximum von Mises Stress (Figure 4.54a)
    - Maximum Resultant Displacement (Figure 4.54b)
    - Total Strain Energy (Figure 4.54c)
    - % Change in Global Criterion (Figure 4.54d)
    - Degrees of Freedom (DOF) (Figure 4.54e)

    The first three criteria for studying the convergence when the p-adaptive method is used generate graphs for the maximum values of the selected property in each iteration loop for the entire model. The fourth option of the set generates a graph for the global criterion of the p-adaptive method. This criterion is specified before running the analysis in the **Options** panel (Figure 4.1a). It can be either the total strain energy, or the RMS of the resultant displacement or

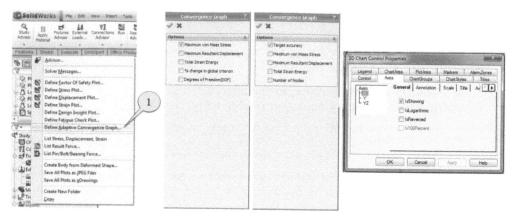

Figure 4.53
Menus for defining a convergence creation of adaptive convergence graphs.

the RMS of von Mises Stress. In our case, the criterion is **Total Energy Strain**. The fifth option generates the graph of the DOFs versus the number of loops. It must be reminded that in the p-adaptive method, the number of DOFs increases (Figure 2.25b).

Optional selections for the **h-adaptive method** are (Figure 4.53c)

- Target accuracy
- Maximum Resultant Displacement
- Total Strain Energy
- % Change in Global Criterion
- Number of Nodes

The unknown convergence graphs that the software generates if the h-adaptive method is selected are **Target accuracy**, which draws the graph of accuracy compared to the target value set at the start of the calculations, and **Number of Nodes**, as it increases throughout the calculations.

- The properties of each of these graphs can be adjusted through the **2D Chart Control Properties** window (Figure 4.53d). Generally, the horizontal axis gives the number of loops and the vertical axis gives the values. If more than one option is selected, the vertical axis is normalised to unity.
- If before running the analysis, the **Trend Tracker** (⊯) is selected, additional graphs of the model properties or simulation results versus the number of loops are generated (Figure 4.55). The shown graphs are for the h-adaptive method with target accuracy of 99% reached within three iterations.
- **Probe graphs** – these are the graph results along any linear entity. The graphs are generated after clicking the **Plot** icon (⊯) from the **Probe** property manager (Figure 4.28). It generates a 2D graph of the preselected results. The horizontal values, which correspond to the locations of nodes along the edge, are normalised to 1. While drawing the graphs, the program assumes linear interpolation between equal distances or between probed locations, or between listed values. For example, the UZ displacements along edge 1 are larger at the head of the chisel and even turn to be negative at the nodes closer to the root (Figure 4.56b). For edge 2, the theory calculates uniform values along the

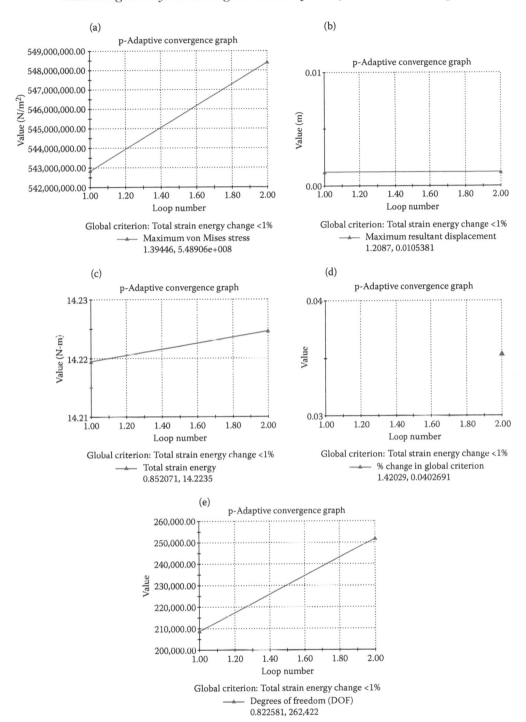

Figure 4.54
Convergence graphs for p-adaptive method. (a) Maximum von Mises stress; (b) maximum resultant displacement; (c) total strain energy; (d) % change in global criterion; (e) degrees of freedom (DOF).

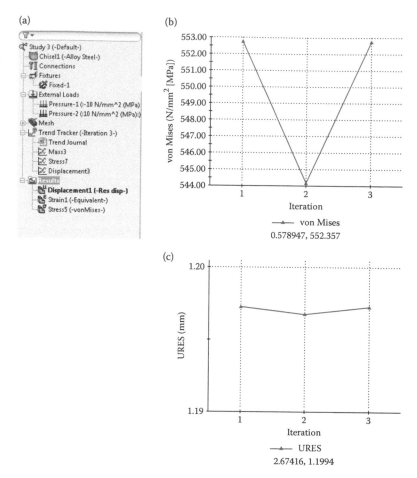

Figure 4.55

Trend Tracker graphs for h-adaptive method. (a) SW Simulation analysis tree with active Trend Tracker; (b) stress Trend Tracker graph; (c) displacement Trend Tracker graph.

edge, while the experiments and our analysis show larger positive values at both ends and smaller displacements at the middle (Figure 4.56c). These values correspond to higher compressive stresses in the ends.

We drew some graphs that can ease the analysis of the results. We furthered our knowledge of **Convergence graphs**, which helps us assess the calculation process. We strengthened our ability to generate **Probe graphs** along an edge for a preselected displacement.

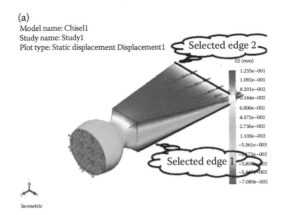

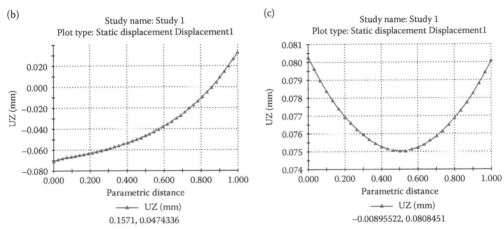

Figure 4.56
Probe UZ displacement graphs. (a) View of the chisel; (b) probe graph of edge 1; (c) probe graph of edge 2.

We learned how to generate graphs:

- **Convergence graphs** – to ease the assessment of the accuracy, computer resources and the speed of the calculations
- **Probe graphs** – to view displacement values along a selected linear entity

---

CHAPTER 5

IMPACT OF MESH DENSITY AND VIEWING MODE ON FINAL RESULTS

5.1 DIFFERENT TYPES OF FEs, REGARDING THE GEOMETRY OF THE MODEL

Based on the geometry of the model, there are a few different types of finite elements (FEs), adopted by SW Simulation – solid FEs, shell FEs and beam FEs. Otherwise, these types of FEs are known as spatial or 3D FEs, plane or 2D FEs and linear or 1D FEs. The program chooses the type of the FE, based on that geometrical criterion automatically, but the user can interact and change software's decision if he/she finds this appropriate. Compared to some other commercial products for FE analysis, SW Simulation does not allow the user to select the type of the FEs. For example, all solid FEs are tetrahedrons (Figure 3.2), and no parallelepipeds or any other prismatic elements can be used. The program automatically creates the following meshes:

- **Solid Mesh** – suitable for bulky objects. The program creates a solid mesh with tetrahedral 3D solid elements for all solid components (Figure 5.1). The user chooses whether linear or parabolic FEs are to be applied.
- **Shell Mesh** – The program automatically creates a shell mesh for shell or plate structures with uniform thicknesses, for example. For sheet metals, the mesh is automatically created at the mid-surface (Figure 5.2a), and the program extracts the shell thickness from the thickness of the sheet metal (Figure 5.2). For other shell structures, there are different options, which can be assigned through the **Shell Definition** property manager and will be discussed later. By default, the top side of the shell is orange and the bottom is grey. This colour legend helps us to analyse easier the directions of the displacements, the positive, and the negative directions of shear stresses, etc. Of course, these colours can be changed (Figure 4.1b).

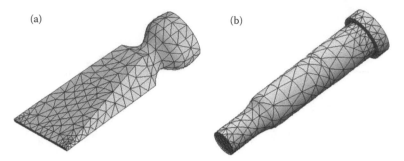

Figure 5.1
Mesh of solid FEs. (a) Solid mesh of a chisel; (b) solid mesh of a hole punch.

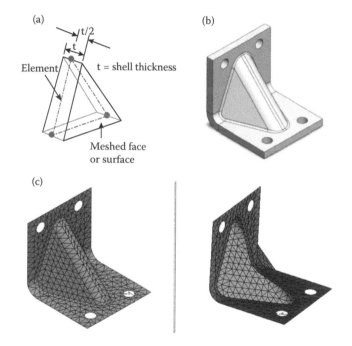

Figure 5.2
Mesh of shell FEs. (a) Sheet metal element (SW Simulation on-line help); (b) CAD model of a connecting part; (c) Shell mesh of a connecting part.

- **Beam Mesh** – The program automatically uses beam mesh for interfering, touching or non-touching within a certain distance of structural elements. For the program, a beam element is a line element defined by two end points (joints) and a cross section. The joints between the elements (the points in magenta for connecting joints and in green for end joints in Figure 5.3) are identified automatically by the program or can be selected manually. There are truss and beam 1D FEs. Beam elements resist to axial, bending, shear and torsional loads, whereas truss elements resist to axial loads only (Figure 5.3).
- **Mixed Mesh** – The program automatically uses a mixed mesh when different geometries are present in the model.

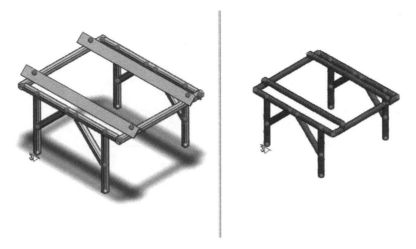

Figure 5.3
CAD model and a mesh of a beam structure.

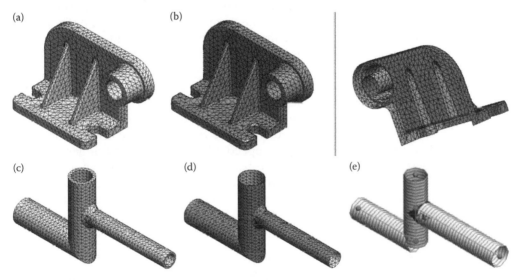

Figure 5.4
CAD models and different types of FE mesh of parts. (a) Solid FE model of a part; (b) shell FE model of the same part – top and bottom views; (c) solid FE model of a welded part; (d) a shell FE model of the same part; (e) a beam FE model of the previous part.

It is important to know that depending on the aim of the analysis, the geometrical model of the structure can be presented through different types of FEs in order to obtain the most appropriate results (Figure 5.4).

We learned about the different types of FEs used by SW Simulation, when meshing a model, in relation to its geometric properties.

We concluded that the user is responsible for selecting the correct type of FEs, yet the software helps them by suggesting the default size and the type of FEs in relation to certain build-in criteria.

We learned that there are three types of FEs in relation to the geometry of the model:

- Solid (3D) FEs – used for meshing bulky models
- Shell (2D) FEs – used for meshing sheet metal-made structures
- Beam (1D) FEs – used for meshing frames, made of structural elements

We know that each model can be meshed using different FEs, yet the final choice depends on the primary objective of the analysis.

5.2 IMPACT OF MESH DENSITY, WHEN STANDARD SOLID MESH IS USED

As was already discussed in Chapter 3, when meshing solid bodies, the program enables the use of two basic types of mesh – standard mesh and curvature-based mesh. As was stated, this property of the elements is closely related to the number of nodes along the element edges and to the power of the polynomial, which transfers calculated node displacements to the displacements along the FE edges. Standard mesh uses FE, in which nodes are situated only in the vertexes of FE (Figure 3.2a). These are **first order** or **linear** elements, because the nodes are connected with straight edges, and the function that describes the displacements within the two vertex nodes (along the FE edge) is a first-order or a linear polynomial.

We will illustrate the impact of the standard mesh on the final results continuing with the analysis of the chisel from the previous chapters.

The first questions to be answered are as follows: How does the density of the mesh influence the computer time and the accuracy of the results? Can we find any optimal ratio or any other criteria, which are to be applied when setting the mesh density?

To answer these questions, we will come back to the first example: an alloy steel chisel, fixed at the root and loaded with two pressure loads – one at the cutting tip and one at the half of the bottom side of the chisel (Figure 2.49). We will discuss the final results from a new point of view.

5.2.1 Coarse Mesh Calculations

At first, let us rename our study from **Study_1** to **Study_Coarse_Mesh**. To do so, we right click on the tab **Study_1** at the bottom bar (Figure 5.5a); then a pop-up menu (Figure 5.5b) opens and we left click on the **Rename** command. Thus, we directly rewrite the new name of the study **Study_Coarse_Mesh** in the tab (Figure 5.5c).

First, we have created a coarse standard mesh with the following properties: draft quality; element size –20 mm; tolerance –1 mm; total nodes –417; total elements –1352; maximum aspect ratio –4.4053; elements with aspect ratio <3 – 98.2%; elements with aspect ratio >10 – 0%; time to complete mesh –0:00:01h. Some plots of the mesh qualities are given in Figure 5.6 and some of the result plots in Figure 5.7.

Values of a few significant results (principal stresses, von Mises stress and displacements) are given in Table 5.1. As has already been said, the final values of some results (stress and strain, for example) depend on the mesh density and the mode (node or element mode). Thus, the two values are provided, and the quantity of the discrepancy is calculated according to the formula $\delta\% = \dfrac{\text{value}_{\text{node mode}} - \text{value}_{\text{element mode}}}{\text{value}_{\text{node mode}}} \times 100\%$. The

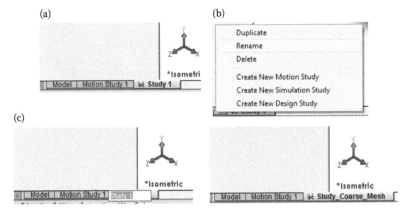

Figure 5.5
Renaming the model. (a) Tab with the old name; (b) pop-up menu; (c) writing the new name directly on the old one.

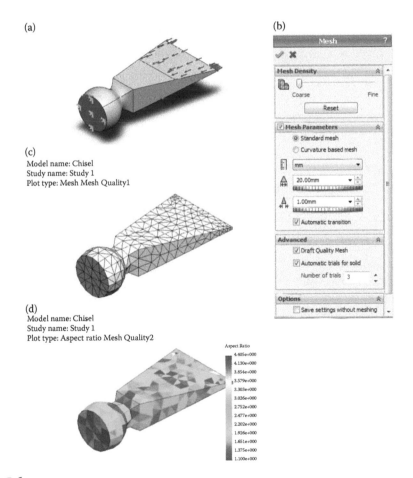

Figure 5.6
Coarse standard mesh model of the chisel. (a) Model of the chisel; (b) Mesh property manager; (c) coarse mesh of the chisel; (d) aspect ratio plot for the created mesh.

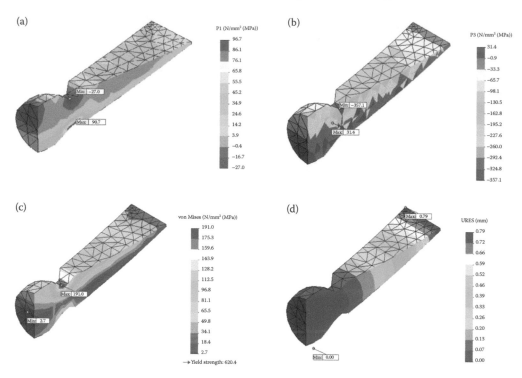

Figure 5.7
Result plots for coarse standard mesh. (a) P1 plot (node mode); (b) P3 plot (element mode); (c) von Mises plot (node mode); (d) displacement plot (UREZ).

Table 5.1
Coarse Standard Mesh Results

	Minimum	δ (%)	Maximum	δ (%)
P1 (node mode) (MPa)	−26.97	−138.0	96.70	−98.0
P1 (element mode) (MPa)	−64.22		191.47	
P3 (node mode) (MPa)	−222.61	−60.4	13.38	−135.1
P3 (element mode) (MPa)	−357.14		31.45	
von Mises (node mode) (MPa)	2.70	−74.1	190.96	−47.5
von Mises (element mode) (MPa)	4.70		281.60	
UZ (mm)	−0.04182		0.09745	
UREZ (mm)	0.000		0.7867	

calculated values show large discrepancies, some of them even larger than 100%. Therefore, it is hard to decide which of the stress values to trust; perhaps element mode values are more appropriate, considering the explanation about how they are calculated.

The conclusion is that the properties of the created mesh are necessary to be reconsidered.

5.2.2   Fine Mesh Calculations

Until now, we have made an analysis using a coarse mesh. Unfortunately, we can hardly rely on the results for there are high discrepancies between the values achieved in the node mode and in the element mode, which is proved by the δ values given in Table 5.1.

Consequently, our next step is to perform the same analysis using a finer mesh. We have to **Create New Simulation Study** (Figure 5.8a) or to **Duplicate** the existing one (Figure 5.8b). If we choose to create a new study, we have to introduce the material, the fixtures and the loads again. Thus, it is chosen to duplicate the existing study. To do so, we right click on the simulation tab at the bottom bar. The tab is titled **Study_Coarse_Mesh**, which is the name of the current study (Figure 5.8c). The **Default Configuration** to use will be kept. Finally, we click **OK**.

The main advantage of the duplicated study is that it has the same properties as the original analysis. Therefore, we will change the mesh properties only. The new mesh is a draft quality mesh (i.e. nodes at the vertexes of the tetrahedral FEs only) as was the first mesh; the element size, which is equal to 2 mm, and the tolerance, which is equal to 0.1 mm, are ten times smaller compared to those of the first mesh (Figure 5.9a and b). Thus, the total number of nodes increases to 107,781 and the total number of elements to 584,340. The maximum aspect ratio is 4.2196, that is, it is smaller compared to the first mesh ratio. The elements with aspect ratio < 3 are 99.9% versus 98.2% for the first mesh, while the elements with aspect ratio > 10 are 0% (Figure 5.9c). The reduction of the aspect ratio and the increase in the percentage of elements with aspect ratio < 3 are the criteria that show the better properties of this newly created mesh. On the contrary, the time to complete the mesh has increased from 0:00:01h to 0:01:13h; the time for running the analysis and the time for showing the plots on the monitor are the same. While running the analysis, we can see that the number of DoFs has also increased to 318,900.

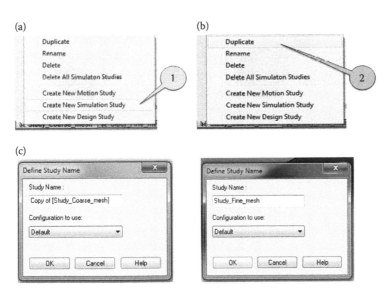

Figure 5.8
How to duplicate an existing study. (a) Start a new study; (b) duplicate an existing study; (c) define the name of the duplicated study.

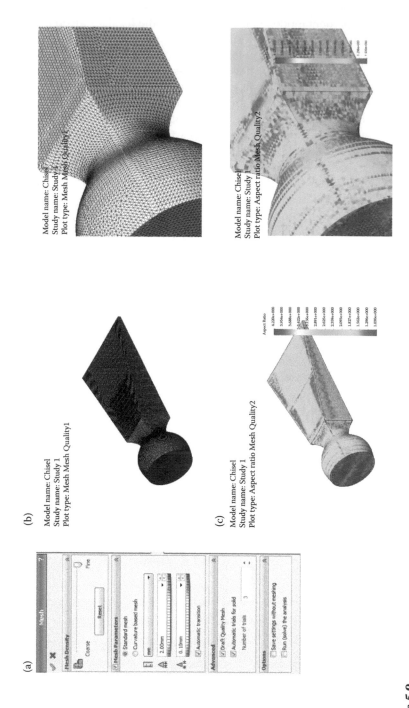

(a)

(b)

Model name: Chisel
Study name: Study 1
Plot type: Mesh Mesh Quality1

Model name: Chisel
Study name: Study 1
Plot type: Mesh Mesh Quality1

(c)

Model name: Chisel
Study name: Study 1
Plot type: Aspect ratio Mesh Quality2

Model name: Chisel
Study name: Study 1
Plot type: Aspect ratio Mesh Quality2

Figure 5.9

Coarse standard mesh model of the chisel. (a) Mesh property manager; (b) fine mesh of the chisel; (c) aspect ratio plot for the created mesh.

While making this exercise, you have to bear in mind that some of the above-mentioned values (total nodes, total elements) can be different from those obtained by your solution depending on the algorithms of creating the mesh. But this is not crucial to the accuracy of the final results.

Thus, before discussing any of the results, we can conclude that we have improved the mesh but we have also increased the computing time.

A few plots of different results are given in Figure 5.10. The outlines of the plotted results are almost similar to the plots from the previous analysis (Figure 5.7), while the values differ significantly.

Table 5.2 summarises the fine mesh results. It is obvious that the discrepancies between the extreme values of the node and the element modes are reasonably reduced. This confirms our previous statement that the finer the mesh, the more accurate the results.

Yet, we cannot provide any quantitative parameter to measure the level of mesh improvement.

5.2.3 Control Mesh Calculations

The version of the controlled mesh combines the advantages of the coarse mesh (quick calculations and minimum required computer resources) with those of the fine mesh (smaller elements, higher accuracy of the results), and as a result, it overcomes their main disadvantages.

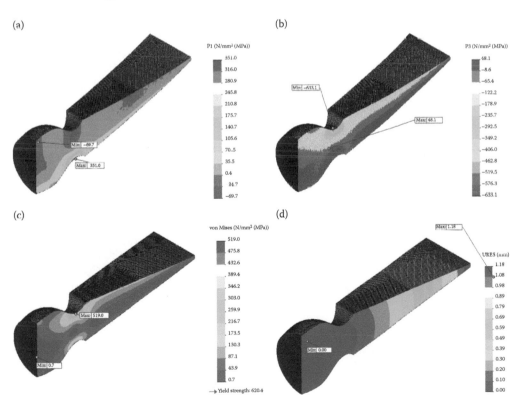

Figure 5.10
Result plots for coarse standard mesh. (a) P1 plot (node mode); (b) P3 plot (element mode); (c) von Mises plot (node mode); (d) displacement plot (UREZ).

Table 5.2
Fine Standard Mesh Results

	Minimum	δ (%)	Maximum	δ (%)
P1 (node mode) (MPa)	−69.72	−21.7	351.04	−13.2
P1 (element mode) (MPa)	−84.86		397.39	
P3 (node mode) (MPa)	−569.81	−11.1	37.27	−29.1
P3 (element mode) (MPa)	−633.09		48.1	
von Mises (node mode) (MPa)	0.700	29.2	519.0	−6.8
von Mises (element mode) (MPa)	0.49522		554.34	
UZ (mm)	−0.070185		0.1246	
UREZ (mm)	0		1.1809	

At first, we duplicate **Study_Coarse_Mesh** according to the way described in Sections 5.2.1 and 5.2.2. The new study is titled **Study_Controlled_Mesh**.

Before starting to define the controlled mesh, it is recommended to make coarse mesh calculations to find where the vulnerable areas of the model are. Thus, we can create a denser mesh around those zones and a coarser mesh all over. As a result, we will have a fine-enough mesh in the vulnerable areas and a comparatively small number of elements/nodes. It is important to remind that the number of nodes is directly related to the number of DoFs, which itself equals the number of the solved equations. Therefore, the smaller this number, the quicker the calculations.

Based on our previous calculations, we know that the vulnerable areas are at the gudgeon of the chisel, that is, these are the areas where the extreme stresses are. Thus, we will reduce the size of elements there and leave the coarse mesh for the other areas. To do so, our first step is to find the narrowest cross section of the chisel, as follows:

1. Finding the intersection of the **Top Plane** and the two side faces of the root loft (Figure 5.11):

Sketch → Convert Entities → Intersection Curve

(a)　　　　　　　　　　　　　　　　　　　　　　　(b)

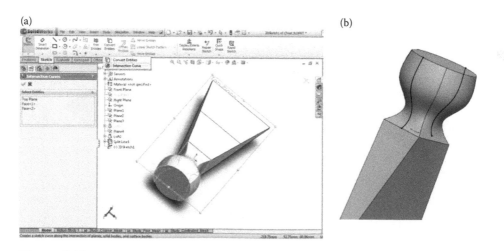

Figure 5.11
Intersection between Top Plane and two side faces of the root. (a) Defining intersection curve; (b) view of intersecting curves.

2. Finding the narrowest section of the root. This is the cross section through the intersection point of the loft curve and the tangent to it (Figure 5.12).
3. Plotting a plane perpendicular to the axis of the chisel and through the newly defined intersection point (Figure 5.13):

Features → Reference Geometry → Plane

4. Defining the split contour – finding the intersection between Plane 5 and the side faces of the root loft (Figure 5.14):

Features → Curves → Split Line

(a) (b)

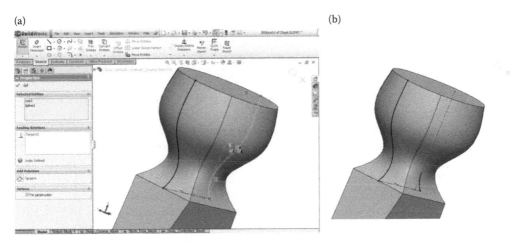

Figure 5.12
Finding the narrowest section of the chisel root – 1. (a) Tangent relation between the construction line and the curve; (b) trimmed to the intersection point curve.

(a) (b)

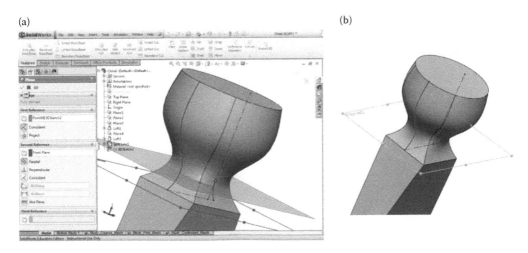

Figure 5.13
Defining the plane of the narrowest section of the chisel root – 2. (a) Definition of a plane through the intersection point; (b) newly defined Plane 5.

139

(a)　　　　　　　　　　　　　　　　　　　　　　　　　　　(b)

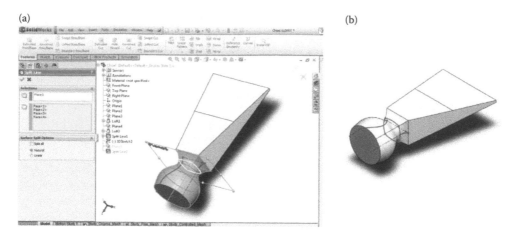

Figure 5.14
Finding the split contour of the narrowest section of the chisel root. (a) Definition of the split line between Plane 5 and the side faces; (b) contour of the split line – magenta line.

Now we are ready to activate the **Mesh Control** command. As we used the **Duplicate** command to start this new study, the properties of the mesh had been copied. Thus, the current mesh consists of elements of size 20 mm and a tolerance of 1 mm, which is the same as those of the mesh of **Study_Coarse_Mesh**.

We will resize the mesh around the three contours shown in Figure 5.15. To do so, we start the **Mesh** pop-up menu from the analysis tree (Figure 3.3b) and left click **Apply Mesh Control** (▣, Figure 3.3b). The **Mesh Control** property manager opens (Figure 5.16). In the blue window **Select Entities**, the signatures of all picked by left clicking in the **Graphics area** entities/edges are displayed. We have chosen to make the mesh finer, targeting the three contours (Figure 5.16a–c). These three groups of edges define three separate **Controls**. The units, the size of the elements and the ratio of decrease are introduced through the **Mesh Property** sub-window. We set **Units** (▣) to be millimetres and **Element size** (⇔) to be 2 mm as is the element size in **Study_Fine_Mesh** and **Ratio** (%), which sets the ratio between the element size in the neighbouring layers to be 1.5. Approximately, this means that the element size of each next layer will be 50% less than the size of the elements from the previous layer. The properties of the entire mesh are kept as they have been set, that is, Units (▣) – mm; Global Size (⇪) – 20 mm; and **Tolerance**

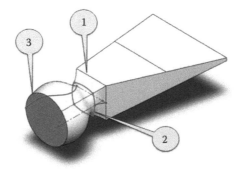

Figure 5.15
Drawing the split contour.

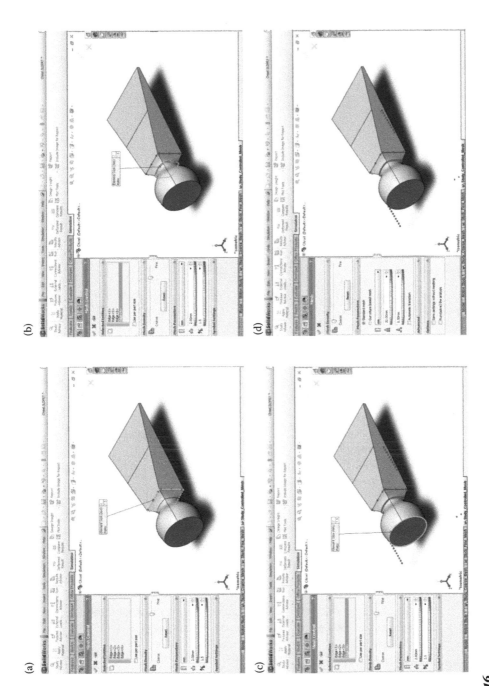

Figure 5.16
Defining the mesh control edges. (a) Defining the controlled edges of the first contour; (b) defining the controlled edges of the second contour; (c) defining the controlled edges of the third contour; (d) properties of the mesh without mesh control.

Figure 5.17
Quality plots of the controlled mesh. (a) Plot of the controlled mesh; (b) aspect ratio of the controlled mesh.

( ) – automatically set by the program to 0.5. The size of the elements equals the size of the elements in **Study_Coarse_Mesh**.

In conclusion, we can say that a combination of the mesh properties of the two previous analyses has been created (Figure 5.17a).

The new mesh consists of 3401 **total nodes** and 14,885 **total elements**, and the **percentage of elements with aspect ratio <3** is 96.7%. The computer time for meshing the model is 0:00:03h. The single worse mesh property is the higher value of the **maximum aspect ratio** (6.8273, Figure 5.17b).

Finally, some results of the calculations with controlled standard mesh are plotted (Figure 5.18), and some results are summarised in Table 5.3 as it is done for both previous case studies.

It is obvious that the extreme, dangerous values are closer to those obtained through fine mesh calculations; some discrepancy values δ still remain high.

5.2.4 Comparison of Results and Conclusions

To enrich the compared data, a fourth case study is set. It repeats the third scenario, but the size of the smallest elements is set to 1 mm (Figure 5.19a), and as a result, the **Tolerance** is reduced to 0.3 compared to 0.5 for the previous mesh. The mesh consists of 7274 nodes and 31,534 elements. It is created for 0:00:07h. The **percentage**

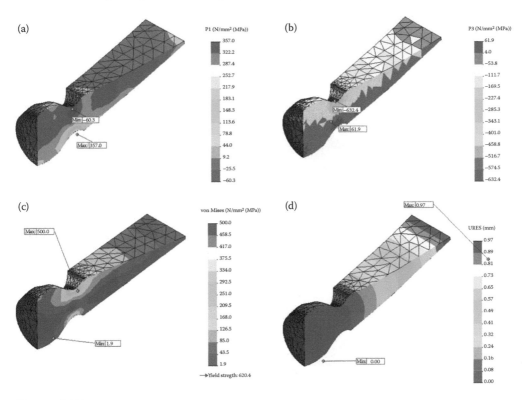

Figure 5.18
Result plots for controlled standard mesh. (a) P1 plot (node mode); (b) P3 plot (element mode); (c) von Mises plot (node mode); (d) displacement plot (UREZ).

Table 5.3
Controlled Standard Mesh Results (Element Size 20 to 2 mm)

	Minimum	δ (%)	Maximum	δ (%)
P1 (node mode) (MPa)	−60.30	−118.4	357.00	−9.9
P1 (element mode) (MPa)	−131.70		392.20	
P3 (node mode) (MPa)	−550.33	−14.9	33.07	−87
P3 (element mode) (MPa)	−632.41		61.89	
von Mises (node mode) (MPa)	1.94	6.7	500.03	−5.6
von Mises (element mode) (MPa)	2.07		527.87	
UZ (mm)	−0.06550		0.114317	
UREZ (mm)	0		0.974802	

of elements with aspect ratio <3 is a little bit smaller (96.4% vs. 96.7%), while the **maximum aspect ratio** is significantly reduced (5.8503 vs. 6.8273; Figure 5.19b).

The results of the four case studies are given in Table 5.4. The trend is as follows: As the size of the elements decreases, the discrepancies between the extreme values for the case studies also decrease. The controlled mesh combines the quick calculations as a direct consequence of the reduced number of nodes with a higher accuracy of the results – the

(a)
Model name: Chisel
Study name: Study_Controlled_Mesh
Plot type: Mesh Mesh Quality1

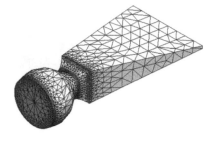

(b)
Model name: Chisel
Study name: Study_Controlled_Mesh_1
Plot type: Aspect ratio Mesh Quality2

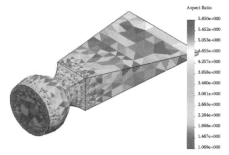

Figure 5.19
Properties of the controlled mesh (1 mm). (a) Plot of the controlled mesh; (b) aspect ratio of the controlled mesh.

discrepancies between the fine mesh results and the controlled mesh results are significantly decreased, compared to the discrepancies between the coarse mesh results.

The last questions to be answered are as follows: Where is the vulnerable area? What are the true extreme values of the stresses and displacements?

No doubt that the vulnerable area is at the gudgeon of the chisel. Despite the chosen mesh density, all stress plots displace the extreme stresses in that area.

The discrepancies between the extreme stress values, regarding the mode of results view, vary significantly versus the mesh density. So, before making conclusions, it is better to filter the results. It is recommended to focus our attention to the positive values of **P1** to get an idea about the vulnerable areas exposed to tension. The maximum **P1** stresses appear in the gudgeon area, and if we exclude the coarse mesh results as inaccurate, we can see that discrepancies of the other sets are smaller than 15%. Even more, we can say that the element mode stress presentations show extreme tensile stresses in the range of 390 to 400 MPa, no matter what the chosen mesh density was. Almost similar is the justification considering the extreme compressive stresses (minimum **P3**). Based on the node mode presentation, their values vary from 550 to 590 MPa, while regarding the element mode view, the compressive stresses are up to 640 MPa. This could be a problem considering that the yield stress of the material is 620 MPa. But if the FEs' size is reduced in the vulnerable area, we see that the compressive stress values reduce to 633.5 MPa for the element mode compared to 603 MPa for the node mode. Thus,

Table 5.4
Compared Results of Different Case Studies

Type of Mesh	Minimum Values				Maximum Values			
	Coarse	Contr. 2 mm	Contr. 1 mm	Fine	Coarse	Contr. 2 mm	Contr. 1 mm	Fine
P1 (node mode) (MPa)	−26.97	−60.30	−74.20	−69.72	96.70	357.00	376.50	351.04
P1 (element mode) (MPa)	−64.22	−131.7	−90.28	−84.86	191.47	392.2	400.45	397.39
Discrepancy δ (%)					−98.00	−9.86	*−6.36*	−13.20
P3 (node mode) (MPa)	−222.61	−550.33	−590.89	−569.81	13.38	33.07	39.41	37.27
P3 (element mode) (MPa)	−357.14	−632.41	−646.35	−633.09	31.45	61.89	55.33	48.1
Discrepancy δ (%)	−60.43	−14.91	*−9.39*	−11.11				
von Mises (node mode) (MPa)	2.70	1.94	1.54	0.700	190.96	500.03	532.14	519.0
von Mises (elem. mode) (MPa)	4.70	2.07	1.22	0.49522	281.60	527.87	559.15	554.34
Discrepancy δ (%)					−47.46	−5.57	*−5.08*	−6.81
UZ (mm)	−0.0418	−0.0655	−0.0666	−0.0702	0.0975	0.1143	0.1153	0.1247
UREZ (mm)					0.7867	0.9748	0.9941	1.1809

depending on the functions of the analysed object, on its importance within the entire production and on its price, the designer decides whether to leave the chisel as it is or to perform some constructive optimisation focusing on the diameter and the curvature of the vulnerable zone. As for the analysed object, the factor of safety is larger than 1, and considering the other criteria, such as the price and the level of reliability and safety of the chisel, the design is preserved and no optimisation activities are planned.

In this section, we solved four case studies based on a standard FE mesh.

We generated a coarse mesh, a fine mesh and both controlled meshes.

Comparing the results and the necessary computer time, it is undoubtedly proved that the controlled mesh is the optimal solution if the controlled entities are correctly selected (Figure. 5.16). The controlled entities can be vertexes, edges or faces. To decide which entities to be controlled, it is recommended to develop a coarse mesh study (Figure 5.7) at first to outline the vulnerable areas.

We learned how to generate different types of standard meshes of solid FEs and compared their advantages and shortcomings. We learned how to create a controlled mesh and defined a criterion on how to choose the controlled entities. The suggested algorithm starts with

- Coarse mesh analysis to outline the vulnerable area.
- Controlled mesh analysis, where the choice of controlled meshed entities is based on the previous stage.
- Second controlled mesh analysis, where the maximal size of the controlled FEs is half the size of the FEs in the previous stage; comparison of both controlled mesh results. This stage is optional.
- Reaching some conclusions about the accuracy of the obtained results and about the necessity of any design optimisation.

5.3 IMPACT OF MESH DENSITY, WHEN CURVATURE-BASED SOLID MESH IS USED

5.3.1 Development of CAD Model of Hole Puncher

The curvature-based mesh is worthy to be used in the FEAs of objects with high curvature values.

Consequently, we have chosen to analyse a hole puncher in this chapter. Its geometry is more complex compared to the geometry of the chisel. There are some areas of high curvature. Thus, the use of curvature-based mesh is preferable.

At first, we will explain how to develop the 3D model of the hole puncher. It consists of three united bodies, each developed using **Sweep** and **Revolve** features. We start with sketching and sweeping the punching/bottom body (Figure 5.20).

The stages to be fulfilled are

1. Sketching the puncher – **Sketch_2** in plane **XY** (Figure 5.20a)
2. Sketching the path – **Sketch_1** in plane **XZ** (Figure 5.20b)
3. Sweeping **Sketch_2** along the path in **Sketch_1** (Figure 5.20c and d)

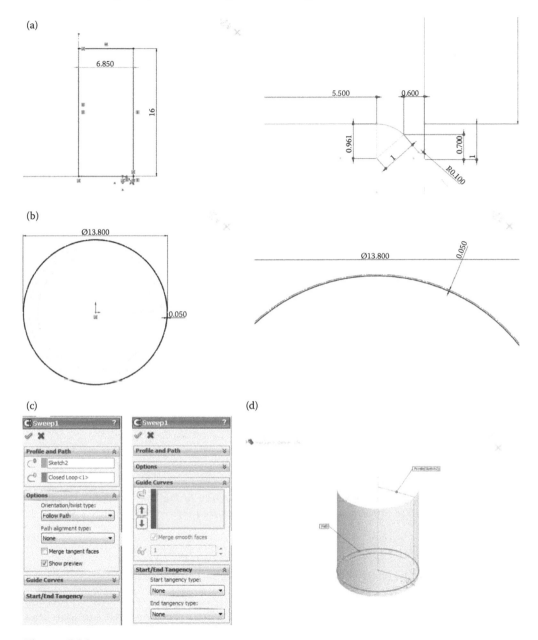

Figure 5.20
Modelling the punching body. (a) Sketching the puncher – Sketch_2 in plane XY; (b) sketching the path – Sketch_1 in plane XZ; (c) options of Sweep property manager; (d) view of the swept sketch and the path.

The next to be modelled is the root of the hole puncher. We start with

1. Sketching the root of the hole puncher – **Sketch_3** in plane **XY** (Figure 5.21a)
2. Revolving **Sketch_3** (Figure 5.21b–d)

The last to be modelled is the mid-body of the hole puncher. To develop its CAD geometry, we will run sketch and revolve functions:

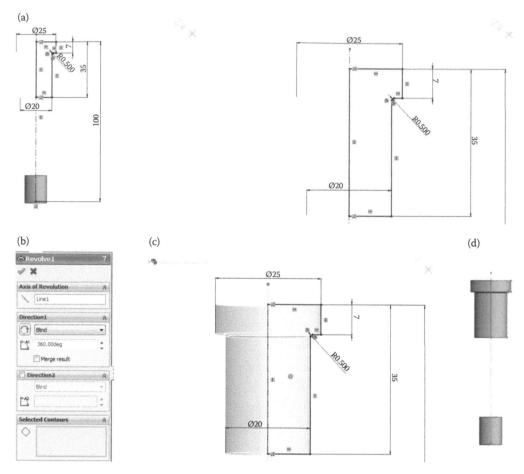

Figure 5.21
Revolving the root of the hole puncher. (a) Sketching the root of the hole puncher – Sketch_3 in plane XY; (b) Revolve property manager; (c) Graphics area view of the revolving Sketch_3; (d) Revolved root.

1. Sketching the mid-body – **Sketch_4** in plane **XY** (Figure 5.22a)
2. Revolving **Sketch_4** (Figure 5.22b–d)

The 3D model developed is shown in Figure 5.23, and therefore, we can start the static analysis.

5.3.2 Development of Hole Puncher Model – Pre-Processor Stage

The first analysis to be done uses coarse mesh and its title is **Study_Coarse_Mesh**.

The material of the body is **alloy steel**. **Alloy steel** is a linear isotropic material with an elastic modulus of 210,000 MPa, Poison's ratio equal to 0.28, mass density of 7700 kg/m^3, tensile/compressive strength of 723.83 MPa and yield strength of 620.42 MPa.

Applied to the model fixtures are **Fixed Geometry** and **Roller/Slider** as shown in Figure 5.24.

(a)

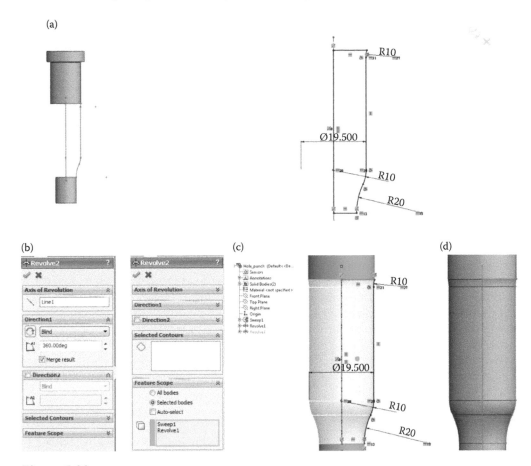

(b)   (c)   (d)

Figure 5.22
Revolving the mid-body of the hole puncher. (a) Sketching the mid-body – Sketch_4 in plane XY;
(b) Revolve property manager; (c) view of the revolving Sketch_4; (d) Revolved mid-body.

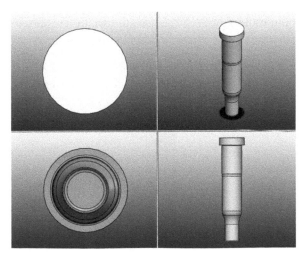

Figure 5.23
Isometric, top, bottom and front views of the modelled hole puncher.

(a)   (b)

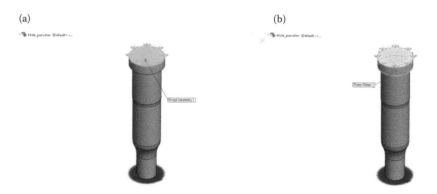

Figure 5.24
Fixtures of the body. (a) Fixed Geometry fixture; (b) Roller/Slider fixture.

The applied loads are **Gravity** (the red arrow) and **Force** (the magenta arrows in Figure 5.25). The gravity load is a volume-distributed load (N/m³), which is equal to the product of the earth's acceleration (9.81 m/s²) and mass density (kg/m³). The force load (equal to 4000 N) is distributed across the face of the fillet of the punching edge load (N/m²). It is calculated as the value of the force divided by the area of the face. This load is parallel to the central axis of the hole puncher.

Regarding the FE model, that is, the mesh properties, we will try to keep the FE size as close as possible to the values assumed in Chapter 4.

5.3.3 Coarse Mesh Calculations

5.3.3.1 Scenario 1 A coarse mesh with the following properties is created (Figure 5.26): **curvature-based mesh** with 4 Jacobian points; Max Element Size –20 mm and Min Element Size –1 mm; Mesh quality – high; total nodes –13,931; total elements –8638; maximum aspect ratio –142.4; percentage of elements with aspect ratio <3 – 92.8%; percentage of elements with aspect ratio >10 – 2.1%; percentage of distorted element (Jacobian) –0%; time to complete mesh –0:00:02h.

Even before running the analysis, it can be said that the created mesh is not very well constructed as the maximum aspect ratio is large and about 2% of the elements

Figure 5.25
Applied loads.

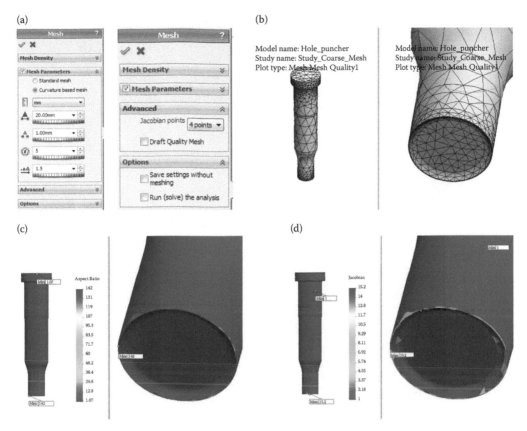

Figure 5.26
Properties of Coarse_Mesh – scenario 1. (a) Mesh property manager; (b) mesh plots; (c) aspect ratio plots; (d) Jacobian plots.

have aspect ratio > 10. Further, these elements can lead to some inaccurately calculated values. It is good that there are no distorted elements. The distorted elements and the elements with a large aspect ratio are a perquisite for lower accuracy of the final results. The aspect ratio, larger than 3, could be a real problem within the calculations.

Some plots and results for **Coarse_Mesh – scenario 1** are shown in Figure 5.27 and in Table 5.5. Discrepancies between the node and the element mode are calculated according to the node mode values, that is, $\delta\% = \dfrac{\text{value}_{\text{node mode}} - \text{value}_{\text{element mode}}}{\text{value}_{\text{node mode}}} \times 100\%$.

5.3.3.2 Scenario 2   Now, a finer, yet, coarse mesh is created (Figure 5.28): curvature-based mesh with 16 Jacobian points; max element size –10 mm and min element size –0.5 mm; total nodes –31,450; total elements –19,749; maximum aspect ratio –154.28 (larger compared to the previous scenario); percentage of elements with aspect ratio <3 – 94.7%; percentage of elements with aspect ratio >10 – 1.07%; percentage of distorted element (Jacobian) –0%; Time to complete mesh –0:00:03h. Despite the rise of the maximum aspect ratio, the percentage of elements with aspect ratio < 3 increases

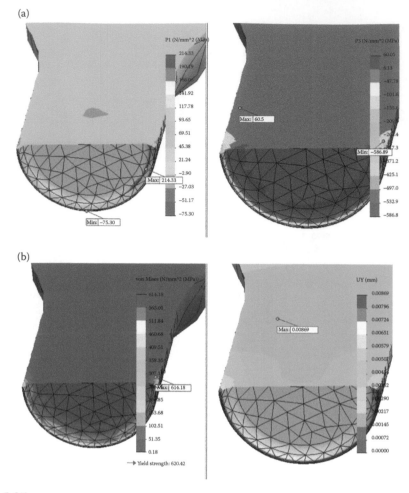

Figure 5.27
Result plots for scenario 1. (a) P1 and P3 stress plots in node mode; (b) von Mises plot in node mode (on the left) and UY displacement (on the right).

Table 5.5
Coarse Mesh Scenario 1 – Results

Extreme Values of Some Stresses and Displacements for Scenario 1			
	Node Mode	Element Mode	Discrepancy δ%
P1 – min value (MPa)	−75.30	−27.21	63.9
P1 – max value (MPa)	214.33	100.31	53.2
P3 – min value (MPa)	−586.89	−502.07	14.5
P3 – max value (MPa)	60.05	13.07	78.2
von Mises – max value (MPa)	614.18	525.26	14.5
UY – max value (mm)	0.00869		
UREZ – max value (mm)	0.00938		

(a)   (b)

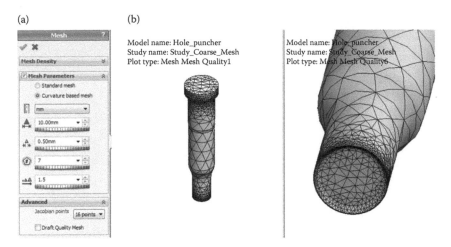

Figure 5.28
Coarse mesh plot for scenario 2. (a) Mesh property manager; (b) mesh plot.

Table 5.6
Coarse Mesh Scenario 2 – Results

Extreme Values of Some Stresses and Displacements for Scenario 2			
	Node Mode	**Element Mode**	**Discrepancy δ%**
P1 – min value (MPa)	−29.66	−13.75	53.6
P1 – max value (MPa)	217.65	108.05	50.4
P3 – min value (MPa)	−635.91	−535.99	15.7
P3 – max value (MPa)	71.49	23.23	67.5
von Mises – max value (MPa)	627.81	554.94	11.6
UY – max value (mm)	0.00881		
UREZ – max value (mm)	0.00950		

and the percentage of those with aspect ratio > 10 decreases; hence, the quality of the mesh is better.

Some results of the run analysis are provided in Table 5.6. They prove that for curvature-based mesh, the decrease in the FE size causes quicker reduction of the discrepancies between the node and the element modes, compared to the same procedure for the standard mesh analysis.

5.3.4   Fine Mesh Calculations

The general size of the second-order elements is assumed to be 2 mm with Jacobian points equal to the maximum provided by the software, that is, 29. Plots of this mesh are given in Figure 5.29. The mesh comprises 131,190 nodes and 85,186 elements. The maximum aspect ratio of 61.128 is related to a single badly configured FE. As a whole, all elements are well configured, and the percentage of those with aspect ratio < 3 is relatively high –98.9%. The percentage of FEs with aspect ratio > 10 is almost zero (0.00822%), and there is one distorted element (0.00117%). The necessary computer time to generate the mesh is 0:00:11h, a little bit longer compared to the two previous scenarios.

(a)   (b)

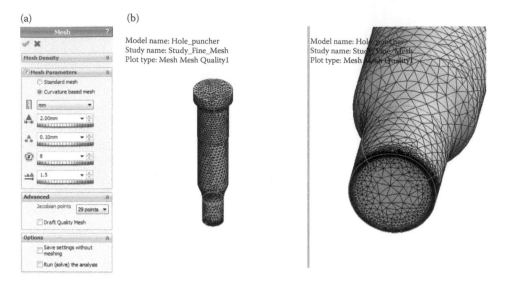

Figure 5.29
Fine mesh plot. (a) Mesh property manager; (b) mesh plot.

While running the analysis, a system of 391,569 equations is being solved for approximately 0:00:09h. Their number equals the number of DoFs of the model. The two principal stress plots, P1 for the highest tensile stresses and P3 for the highest compressive stresses, are shown in Figure 5.30, and the compared results are given in Table 5.7.

Contrary to our expectations, the discrepancies δ do not decrease. We will try to explain that fact, a few pages later, when we compare the standard mesh results to the curvature-based mesh results. Now it is enough to know that the observed consistency in the results is partially due to the geometry of the analysed part and partially due to the fact that the curvature-based mesh provides high accuracy even for FE of a larger size.

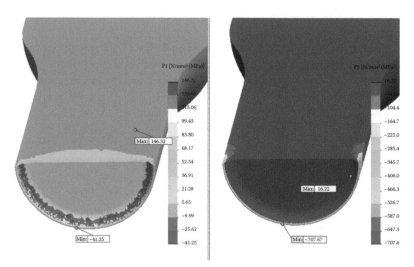

Figure 5.30
Principal stress plot – element mode.

Table 5.7
Fine Mesh Results

Extreme Values of Some Stresses and Displacements for Fine_Mesh Scenario			
	Node Mode	Element Mode	Discrepancy δ%
P1 – min value (MPa)	−89.10	−41.25	53.7
P1 – max value (MPa)	353.79	146.32	58.6
P3 – min value (MPa)	−878.56	−707.67	19.5
P3 – max value (MPa)	37.26	16.22	56.5
von Mises – max value (MPa)	964.45	647.98	32.8
UY – max value (mm)	0.00893		
UREZ – max value (mm)	0.00965		

As conclusion, it is enough to remember that the decrease in the element size is not a panacea to all problems regarding the model and does not always guarantee a quick convergence to the accurate values.

5.3.5 *Control Mesh Calculations*

It is obvious that the most vulnerable area is the cutting edge of the hole puncher. Thus, it has been decided to decrease the element size of all FEs that are in its neighbourhood. Consequently, the control mesh will be applied to the cutting edge surroundings.

At first, we create a split line around the punching body (Figure 5.31). It duplicates **Sketch_1** (Figure 5.31a). The easiest way to create the split line is to open the **Split Line** property manager (Figure 5.31b):

Features → Curves → Split Line (⬚)

and to define the split line as an intersection of the **Top Plane** and the outer face of the punching body. The result is a new entity **Split Line3** (Figure 5.31c).

The next step is to generate mesh control properties. Two scenarios are designed.

5.3.5.1 Scenario 3 This scenario uses a single **Mesh Control** option (Figure 5.32). The size of the elements of the blue-coloured faces (Figure 5.32a) is set to 0.5 mm, while the decreased ratio of two neighbouring element layers is 1.5. The maximum size of the rest of the FEs is 5 mm, and the minimum size of FEs is 0.5 mm. Jacobian points is 29 (Figure 5.32b).

The new curvature-based mesh consists of 190,797 nodes and 122,089 FEs. The maximum aspect ratio is 8.3045 (Figure 5.33b), and it is significantly reduced compared to the values of the previous scenarios. There are no badly configured elements – either distorted or elements with aspect ratio > 10, and the percentage of FE with an aspect ratio < 3 is the highest of all considered scenarios (99.2%). The time for mesh creation is also short (0:00:17h), because of the comparatively small controlled areas (Figure 5.33a).

The run analyses solve a model of 571,944 DoFs within 21 s.

A plot of the two extreme principal stresses, P1 and P3, is given in Figure 5.34. The numerical values of the results are in Table 5.8.

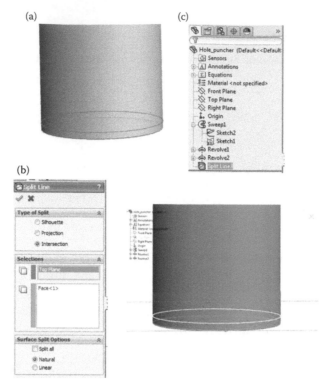

Figure 5.31
Generating Split Line around the punching body. (a) View of Sketch_1; (b) Split Line property manager and graphic area view; (c) model tree.

Figure 5.32
Defining mesh control and mesh properties. (a) Mesh Control property manager; (b) Mesh property manager.

(a)   (b)

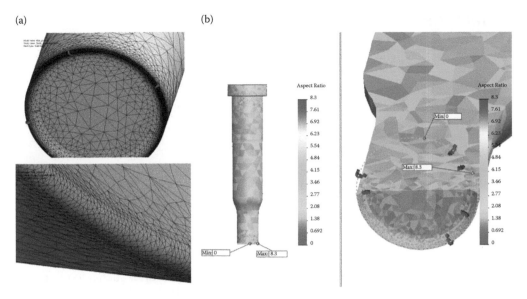

Figure 5.33
(a) Controlled mesh and (b) aspect ratio plots.

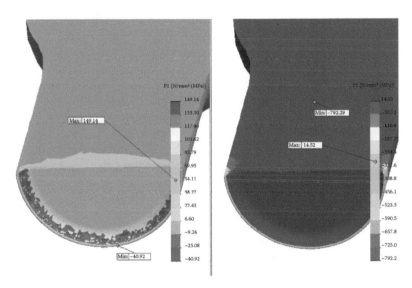

Figure 5.34
Principal stress plot – controlled mesh, element mode, scenario 3.

Compared to the fine mesh case, some of the discrepancies between the two modes are significantly reduced, while the solution time and the use of computer resources increase a little.

Unfortunately, there are two values that grasp our attention. These are the extreme compressive stresses P3 and the maximum von Mises stress. Both values are interrelated as the von Mises stresses are calculated, regarding the P3 values

$(\sigma_e = \sqrt{\dfrac{(\sigma_1 - \sigma_2)^2 + (\sigma_2 - \sigma_3)^2 + (\sigma_3 - \sigma_1)^2}{2}}$, see p. 104). The calculated values exceed

Table 5.8
Controlled Mesh Results – Scenario 3

Extreme Values of Some Stresses and Displacements for Scenario 3			
	Node Mode	Element Mode	Discrepancy δ%
P1 – min value (MPa)	−55.59	−40.92	26.4
P1 – max value (MPa)	179.93	149.14	17.1
P3 – min value (MPa)	−1005.80	−792.29	21.2
P3 – max value (MPa)	30.90	14.52	53.0
von Mises – max value (MPa)	891.88	727.49	18.4
UY – max value (mm)	0.00893		
UREZ – max value (mm)	0.00965		

even the compressive strength of the material. If we have measured the stresses in a physical model of the hole puncher, we would have observed two facts:

- The model does not destroy and is ready to operate even after such a loading; only some really small areas of the cutting edge are deformed.
- The highest measure compressive stresses are much smaller than the values calculated through FE analysis.

The explanation of that fact is quite simple if we consider the nature of both processes. While developing a FE model and further calculating the results, we do not take into consideration that the material itself transfers the loads across small areas, redistributes them more evenly across these areas and thus reduces the extreme stresses. The software is unable to do so; it calculates the stresses at the very node or FE. We must remind that the minimum FE size of the mesh is 0.5 mm, while the radius of the cutting edge is 0.1 mm (Figure 5.20a), which also reduces the accuracy of the FE model around the cutting edge area. The extreme compressive stresses appear in a very thin inconsistent strip at the outer face of the punching body. This allows us to assume that these large compressive stresses are a consequence of the calculating algorithms and the method's assumptions.

5.3.5.2 Scenario 4   The program builds the controlled mesh based on four different **mesh controls**.

The first **control** is related to the face of the cutting edge (Figure 5.35a) and sets the FE size to 0.1 mm, a value that equals the radius of the edge (Figure 5.20a). The second **control doubles** the FE size in the two neighbouring rings (Figure 5.35b). The third **control** increases the FE size to 0.3 mm for the second inner ring (Figure 5.35c), and the last **control** defines the FE size of the bottom face of the hole puncher to be 0.5 mm (Figure 5.35d).

The properties of the entire mesh are kept as they are in scenario 3; only the maximum FE size is reduced to 2 mm, and the minimum FE size is 0.4 mm (Figure 5.36a).

The new mesh possesses the following details: total nodes –231,594; total elements 150,693; max element size –2 mm; min element size –0.4 mm; maximum aspect ratio –8.526; percentage of elements with aspect ratio < 3 – 99.4%; % of distorted elements –0; time to complete mesh –0:00:23h (Figure 5.36b and c).

Running the analysis solves a FE model of 692,769 DoFs and takes about 25 s.

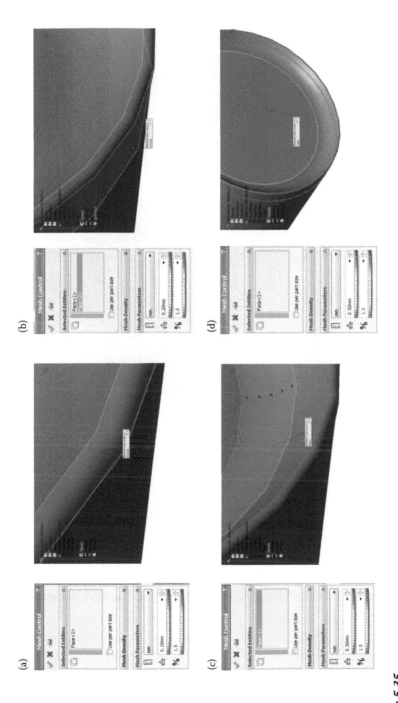

Figure 5.35
Different controlled options. Controlled options for (a) Control 1; (b) Control 2; (c) Control 3; (d) Control 4.

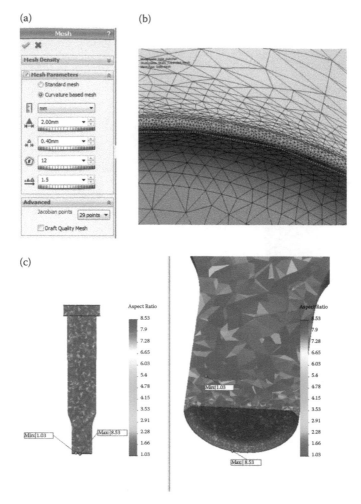

Figure 5.36
Mesh for scenario 4. (a) Mesh property manager; (b) mesh plot – controlled area; (c) aspect ratio plot.

The **SW Simulation analysis tree** shows the four mesh controls and the final plots (Figure 5.37a). The principal stress plots are given in Figure 5.37b. Some of the results are presented in Table 5.9.

5.3.6 Comparison of Results and Conclusions for Curvature-Based Mesh

Results of these five case studies are summarised in Tables 5.10 and 5.11. The trend of quicker convergence to the accurate results using comparatively limited computer resources when controlled mesh is used is even stronger in comparison to the standard mesh.

Yet the use of elements with a high **aspect ratio** sharply decreases the level of accuracy. The existence of distorted elements and the relatively small percentage of FEs with aspect ratio < 3 (see **Fine_Mesh** study) also make worse the quality of the created mesh. One of the possible solutions of that problem is the use of controlled

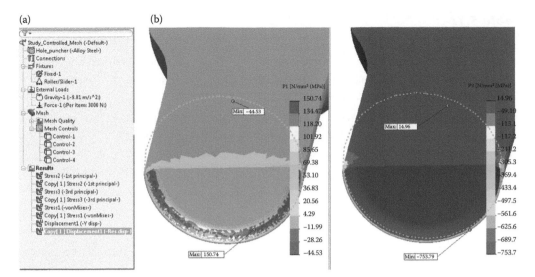

Figure 5.37
Results for scenario 4. (a) SW Simulation tree; (b) principal stress plots.

Table 5.9
Controlled Mesh Results – Scenario 4

	Extreme Values of Some Stresses and Displacements for Scenario 4		
	Node Mode	Element Mode	Discrepancy δ%
P1 – min value (MPa)	−53.35	−44.53	16.5
P1 – max value (MPa)	181.94	150.74	17.1
P3 – min value (MPa)	−926.22	−753.79	18.6
P3 – max value (MPa)	32.60	14.96	54.1
von Mises – max value (MPa)	858.10	696.28	18.9
UY – max value (mm)	0.00893		
UREZ – max value (mm)	0.00965		

Table 5.10
Extreme Values of Some Stresses and Displacements at Node Mode

	Coarse_Mesh Scenarios		Fine_Mesh Scenario	Controlled_Mesh Scenarios	
	Scenario 1	Scenario 2	Scenario	Scenario 3	Scenario 4
P1 – min value (MPa)	−75.30	−29.66	−89.10	−55.59	−53.35
P1 – max value (MPa)	214.33	217.65	353.79	179.93	181.94
P3 – min value (MPa)	−586.89	−635.91	−878.56	−1005.80	−926.22
P3 – max value (MPa)	60.05	71.49	37.26	30.90	32.60
von Mises – max value (MPa)	614.18	627.81	964.45	891.88	858.10
UY – max value (mm)	0.00869	0.00881	0.00893	0.00893	0.00893
UREZ – max value (mm)	0.00938	0.00950	0.00965	0.00965	0.00965

Table 5.11

Extreme Values of Some Stresses and Displacements at Element Mode

| | Coarse_Mesh Scenarios | | Fine_Mesh Scenario | Controlled_Mesh Scenarios | |
	Scenario 1	Scenario 2		Scenario 3	Scenario 4
P1 – min value (MPa)	−27.21	−13.75	−41.25	−40.92	−44.53
P1 – max value (MPa)	100.31	108.05	146.32	149.14	150.74
P3 – min value (MPa)	−502.07	−535.99	−707.67	−792.29	−753.79
P3 – max value (MPa)	13.07	23.23	16.22	14.52	14.96
von Mises – max value (MPa)	525.26	554.94	647.98	727.49	696.28

mesh, where the controlled entities are selected in the neighbourhood of the badly configured FEs' areas.

Another interesting question that is yet to be answered treats the high stress values of min P3 and max von Mises in the node mode for the **Fine_Mesh** and **Controlled_Mesh** scenarios. It is obvious that values that are far beyond the yielding stress (620.42 MPa) and the tensile/compressive strength (723.83 MPa) are forbidden. Yet they exist, and even more, we do accept this phenomenon. This can be explained as follows:

1. These are case studies and we do not focus on achieving a certain factor of safety. But is this explanation convincing enough?
2. You can see that the extreme values appear and impact only very limited vulnerable zones. These high stresses do not spread around as it is in the real material, and the discrepancy between the extreme node values and the extreme element values is really high. The software calculates the element values in relation to all node values related to the calculated FE. Thus, if there is even a one large node value, it reflects the element value, and there is no way to consider that the rest of the node values are far below the yielding stresses. Yet such a large range of node values related to a single FE justifies the observed high discrepancies between the element mode and the node mode. Thus, the discussed extreme values are much more a consequence of our calculations than of the physical behaviour of the analysed object.
3. Another proof of our previous statement is the following fact: these values do not exceed the yielding strength for the first two cases, particularly for the element mode. Thus, they are strongly impacted by the chosen mesh properties, that is, it is time for the next question.

What is the optimal ratio between the mesh properties and the accuracy of the results? Unfortunately, there is no precise universal answer to that question, and I cannot provide here a brief instruction on the topic. The right choice is a matter of understanding and experience and a certain degree of intuition, which combines the designer's practical and theoretical knowledge on the function of the structures.

We solved five case studies using a curvature-based mesh in this section.

Undoubtedly, we proved the advantages of the controlled mesh by using more than one mesh control.

We proved that the decrease in the element size is not the only possible and right way to achieve an accurate solution.

We focused our attention on some guidelines how to recognise the numerical spots, due to wrong modelling or analysis faults, from the real errors, based on wrong model development or misunderstanding of model operation.

We learned how to generate different types of curvature-based meshes.

We learned how to create a controlled mesh with more than one mesh control.

We discussed some guidelines how to pick out the numerical spots from the errors due to wrongly developed models and analysis.

5.4 IMPACT OF MESH DENSITY ON CALCULATION TIME AND ACCURACY

Now, it is time to discuss about the way in which mesh density influences the calculation time. It is important to know that this characteristic is rather partial as it is strongly influenced by the computer configuration – the more powerful the computer, the shorter that time.

The basic properties of all case studies for standard and for curvature-based meshes are systematised in Tables 5.12 and 5.13.

Table 5.12
Mesh Properties for Standard Mesh

	Coarse_Mesh Scenario	Fine_Mesh Scenario	Controlled_Mesh Scenarios	
			Scenario 1	Scenario 2
Global Average Element Size (mm)	20	2	20	20
Tolerance (mm)	1	0.1	0.5	0.3
Max Controlled FE Size (mm)			2	1
Ratio between Neighbouring Layers			1.5	1.5
Total Nodes	417	107,781	3401	7274
Total Elements	1352	584,340	14,885	31,534
Maximum Aspect Ratio	4.4053	4.2196	6.8273	5.8503
Percentage of Elements with Aspect Ratio <3 (%)	98.2	99.9	96.7	96.4
Percentage of Elements with Aspect Ratio >10 (%)	0	0	0	0
Time to Complete Mesh (s)	1	73	3	7

Table 5.13
Mesh Properties for Curvature-Based Mesh

	Coarse_Mesh Scenarios		Fine_Mesh Scenario	Controlled_Mesh Scenarios	
	Scenario 1	Scenario 2	Scenario	Scenario 3	Scenario 4
Jacobian Points	4	16	29	29	29
Max Element Size (mm)	20	10	2	5	2
Min Element Size (mm)	1	0.5	0.1	0.5	0.4
Min Number of Elements in a Circle	5	7	8	12	12
Element Size Growth Ratio	1.5	1.5	1.5	1.5	1.5
Max Controlled FE Size (mm)				0.5	0.1 ÷ 0.5
Ratio between Neighbouring Layers				1.5	1.5
Total Nodes	13,931	31,450	131,190	190,797	231,594
Total Elements	8639	19,749	85,186	122,089	150,693
Maximum Aspect Ratio	142.4	154.28	61.128	8.3045	8.526
Percentage of Elements with Aspect Ratio <3 (%)	92.8	94.7	98.9	99.2	99.4
Percentage of Elements with Aspect Ratio >10 (%)	2.10	1.07	0.00822	0	0
% of Distorted Elements (Jacobian) (%)	0	0	0.0011711	0	0
Time to Complete Mesh (s)	2	3	11	17	23
DoFs			391,569	571,994	692,769
Time to Solve the System(s)	1	2	9	21	25

If we compare the calculation time for one and the same computer, we can make the following guiding conclusions:

- The denser the mesh is, the longer the calculations last, that is, the denser mesh is a perquisite for a larger total number of nodes, and consequently a larger number of DoFs and a larger number of solved equations.
- The standard mesh uses FEs of the first order, while the curvature-based mesh uses FEs of the second order. Thus, the total number of nodes, respectively, DoFs, corresponding to the equal total number of FEs for the standard mesh is smaller compared to those for the curvature-based mesh. Therefore, based on a relatively equal number of FEs, the standard mesh provides quicker solution, that is, uses smaller computer resources and time.
- Based on the higher order of the used FEs, the curvature-based mesh provides quicker convergence to the accurate results, as far as the mesh density is concerned.

Table 5.14
Extreme Stress Values' Discrepancies for Standard Mesh Calculations (%)

	Coarse_Mesh Scenarios	Fine_Mesh Scenario	Controlled_Mesh Scenarios	
			Scenario 1	Scenario 2
P1 – Min Value	−138.0	−21.7	−118.4	−21.7
P1 – Max Value	98.0	−13.2	−9.9	−6.4
P3 – Min Value	−60.4	−11.1	−14.9	−9.4
P3 – Max Value	−135.1	−29.1	−87.0	−40.4
von Mises – Max Value	−47.5	−6.8	−5.6	−5.07

Table 5.15
Extreme Stress Values' Discrepancies for Curvature-Based Mesh Calculations (%)

	Coarse_Mesh Scenarios		Fine_Mesh Scenario	Controlled_Mesh Scenarios	
	Scenario 1	Scenario 2		Scenario 3	Scenario 4
P1 – Min Value	63.9	53.6	53.7	26.4	16.5
P1 – Max Value	53.2	50.4	58.6	17.1	17.1
P3 – Min Value	14.5	15.7	19.5	21.2	18.6
P3 – Max Value	78.2	67.5	56.5	53.0	54.1
von Mises – Max Value	14.5	11.6	32.8	18.4	18.9

5.5 COMPARISON BETWEEN THE NODE MODE AND THE ELEMENT MODE

We already know that the mode selection (a node mode or an element mode) impacts only the presented results. The type of the chosen mode does not influence the calculation flow and accuracy of the results. In fact, this is a choice that affects the final presentation and the report. Based on what we have discussed in Sections 5.2 and 5.3, we can conclude that the element mode is better for presenting numerical values, graphs and plots. Additionally, based on the used techniques for the calculation of element values, that mode cuts the extreme picks that can appear in the node mode.

As the magnitudes/absolute values of the analysed properties become larger, the discrepancies between the node mode and the element mode become lower. The decrease in those discrepancies is strongly influenced by the absolute values of the compared properties than the density and the type of the mesh (Tables 5.14 and 5.15).

5.6 FINAL RECOMMENDATIONS ON SELECTION OF MESH TYPE

Here are some guidelines on how to choose the type and the properties of the mesh to be sure that your analysis is correct:

- Start with a standard mesh, accepting the properties the program suggests.
- If there is a high percentage of elements with an aspect ratio > 10 or a low percentage of elements with an aspect ratio < 3, try either to reduce the FE size or to change the type of the mesh to a curvature-based one. The second solution is recommended if you have a model with a complex geometry or if there are distorted (Jacobian) elements.
- Finally, as the software calculates the displacements first, the stresses are based on those data. Therefore, it is harder to achieve a high level for accuracy for the stresses than for the displacement (see the extreme displacement values for Fine_Mesh and Controlled_Mesh scenarios for the curvature-based mesh, which are equal but generate different extreme stresses).
- Always pay attention to the origin of the extreme stress values, and before continuing, try to explain their origin: are they due to the numerical algorithms and introduced mesh properties or due to mistakes in the model development and the consequent analysis?

We tried to compare the properties of standard and curvature-based meshes and write guidelines on how to start the solution, as far as the mesh creation is concerned.

Additionally, some recommendations on processing the analysis if the obtained results do not coincide with our expectations are given.

We learned how to assess the quality of the created mesh based on the final results and how to improve the accuracy of the results by changing the mesh properties.

CHAPTER 6

STATIC ANALYSIS OF SOLID BODY WITH CIRCULAR OR PLANAR SYMMETRY

6.1 DEVELOPMENT OF CAD MODELS OF THE ANALYSED BODIES

6.1.1 Geometrical Model of a Body with Circular Symmetry

A CAD model of a wheel will be developed. At first, we will develop the entire wheel without considering any simplifications due to existing circular symmetry. Further, we will cut out only one repeating pattern, and finally, we will make a static study on the cut pattern.

All stages of geometrical modelling of the object are briefly outlined:

1. Start **a new model**:

$$\text{File} \rightarrow \text{New} \rightarrow \text{Part} \rightarrow \text{OK}$$

   Save file as **Wheel. sldprt**.
2. Define the used units – **System SI** of millimetre, gram and second units is selected.

$$\text{Tools} \rightarrow \text{Options} \rightarrow \text{Document Properties} \rightarrow \text{Units} \rightarrow \text{MMGS} \rightarrow \text{OK}$$

3. Draw the first sketch. This is a circle with a diameter of 80 mm, drawn in the **Front Plane**

$$\text{Sketch} \rightarrow \text{Front Plane} \rightarrow \text{Circle} \rightarrow \text{OK}$$

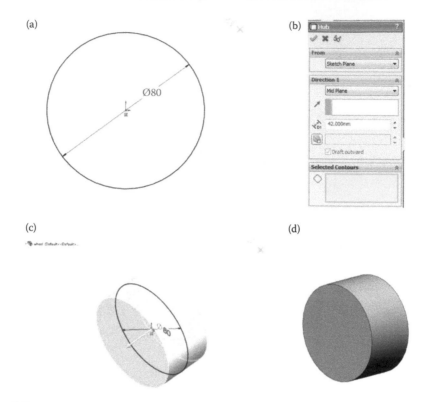

Figure 6.1
Development of the CAD model of the wheel – stages 3 and 4. (a) Sketch1 in Front Plane. (b) Hub Extrude property manager. (c) Graphic areas view while Extrude property manager is active. (d) Extruded Hub component.

Automatically the sketch is indexed as **Sketch1** by the program (Figure 6.1a).

4. Extrude **Sketch1** to feature the hub (Figure 6.1):

$$Features \rightarrow Extrude\ Boss/Base \rightarrow OK$$

The options of that **Extrude Boss/Base** command (⬛) are shown in Figure 6.1b and are given as follows – <u>From:</u> Sketch Plane; <u>Direction 1:</u> Condition type Mid Plane; Direction of Extrusion (↗), nothing is picked as the direction is normal to the sketch plane; Depth (↕), 42 mm; <u>Selected Contours</u> (◇), no contour is picked as the entire sketch will be extruded.

5. Draw the second sketch in the **Right Plane** (Figure 6.2a):

$$Sketch \rightarrow Right\ Plane \rightarrow Sketch2 \rightarrow OK$$

6. Revolve **Sketch2** and make the rim and web (Figure 6.2):

$$Features \rightarrow Revolve\ Boss/Base \rightarrow OK$$

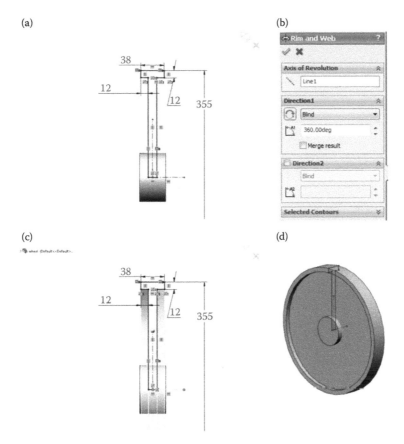

Figure 6.2
Development of the CAD model of the wheel – stages 5 and 6. (a) Sketch2 in Front Plane. (b) Rim and Web Revolve property manager. (c) Graphic areas view while Revolve property manager is active. (d) Extruded Rim and Web component.

The options of the **Revolve Boss/Base** command (⊕) are shown in Figure 6.2b and are given as follows – <u>Axis of Revolution</u> (↘): Line 1; <u>Direction 1:</u> Revolve type (⊙), Blind to revolve the sketch in one direction; Angle of rotation (↥) – 360.00 deg.

7. Draw the third sketch in the **Top Plane** (Figure 6.3a):

Sketch → Top Plane → Sketch3 → OK

8. Extrude **Sketch3** to feature the rib (Figure 6.3):

Features → Extrude Boss/Base → OK

The options of the **Extrude Boss/Base** command (⬛) are shown in Figure 6.3b and are given as follows – <u>From:</u> Sketch Plane; <u>Direction 1:</u> Condition type – Mid Plane; Depth (↧) – 6 mm.

169

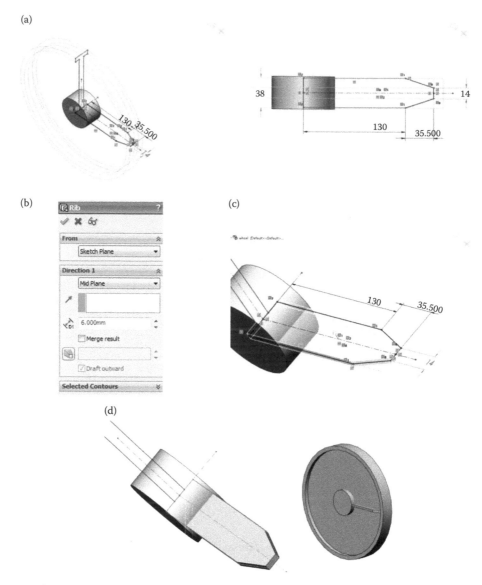

Figure 6.3
Development of the CAD model of the wheel – stages 7 and 8. (a) Sketch3 in Top Plane. (b) Rib Extrude property manager. (c) Graphic areas view while Extrude property manager is active. (d) Extruded Rib component.

9. Draw the fourth sketch in the **Front Plane** (Figure 6.4a):

Sketch → Front Plane → Sketch4 → OK

The sketch is a circle with a diameter of 38 mm.
10. Extrude **Sketch4** to feature the cut through the hub (Figure 6.4b–d):

Features → Extruded Cut → OK

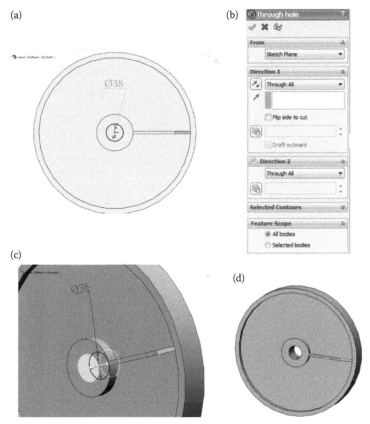

Figure 6.4
*Development of the CAD model of the wheel – stages 9 and 10. (a) Sketch4 in Front Plane.
(b) Through hole extruded cut property manager. (c) Graphic areas view while Extruded Cut
property manager is active. (d) Cut Through hole component.*

The options of **Extruded Cut** (▣) are From: Sketch Plane; Direction 1: End
condition type – Through all; Direction 2: End condition type – Through all;
Feature Scope: All bodies.

11. Draw the fifth sketch in the **Front Plane** (Figure 6.5a):

Sketch → Front Plane → Sketch5 → OK

12. Extrude **Sketch5** to feature the cut through the wheel (Figure 6.5b–d):

Features → Extruded Cut → OK

The options of **Extruded Cut** (▣) are From: Sketch Plane; Direction 1: End
condition type – Through all; Direction 2: End condition type – Through all;
Feature Scope: All bodies.

This stage is similar to stage 10. As a result, the **Lightning hole** component
is created.

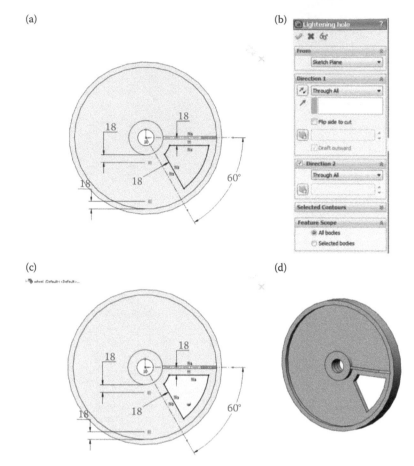

Figure 6.5
Development of the CAD model of the wheel – stages 11 and 12. (a) Sketch5 in Front Plane. (b) Lightning hole extruded cut property manager. (c) Graphic areas view while Extruded Cut property manager is active. (d) Cut Lightning hole component.

13. Feature the fillets at the edges of the lightning hole (Figure 6.6a–c):

Features → Fillet → OK

The options of the **Fillet** property manager (⬚) are Items to Fillet: Constant radius (⌒) – 9 mm; Items to fillet (⬚) – pick the four edges of the lightning hole; Full preview – checked; Setback parameters: no data are entered; Fillet Options: Select through faces – checked; Overflow type – Default.

14. Defining the central axis of the model – **Axis 1**:

Features → Reference Geometry → Axis → OK

The new axis will be defined as the axis of a cylindrical face (Figure 6.6e and f).

The options of the **Axis** property manager (⬚) are Reference entity (⬚) – Face 1 (see Figure 6.6d). This is the guiding **Cylindrical/Conical Face** (⬚).

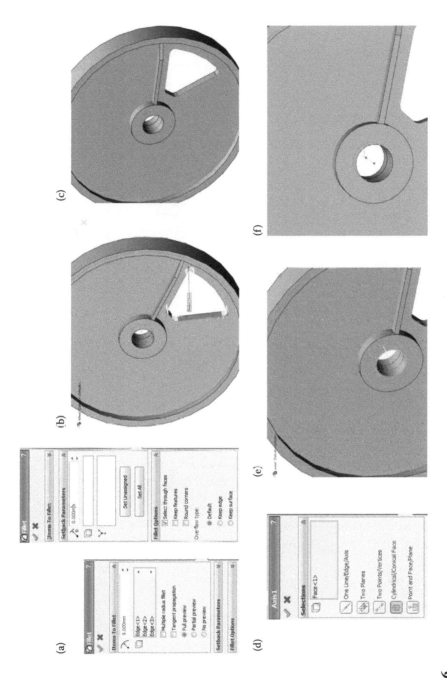

Figure 6.6

Development of the CAD model of the wheel – stages 13 and 14. (a) Fillet property manager. (b) Selected edges preview. (c) Filleted Lightning hole component. (d) Axis property manager. (e) Selected reference entity. (f) Defined new axis – Axis 1.

15. Copying some of the developed components using the **Circular Pattern** command:

$$\text{Features} \rightarrow \text{CirPattern} \rightarrow \text{OK}$$

The options of the **CirPattern** property manager (⊞) are defined as follows (Figure 6.7a) – <u>Parameters</u>: Axis of rotation (⊙) – Axis 1; Angle (⟲) – sets the angle between the instances. It is equal to (360 deg/to the angle of the Lightning hole),

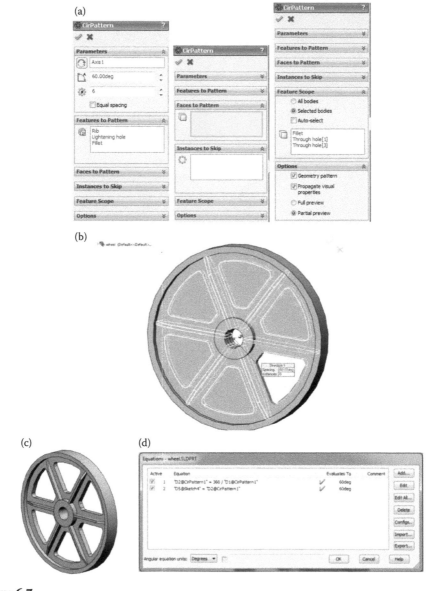

Figure 6.7
Development of the CAD model of the wheel – stages 15 and 16. (a) CirPattern property manager. (b) Graphic areas view while CirPattern property manager is active. (c) Patterned entities. (d) Defined equations.

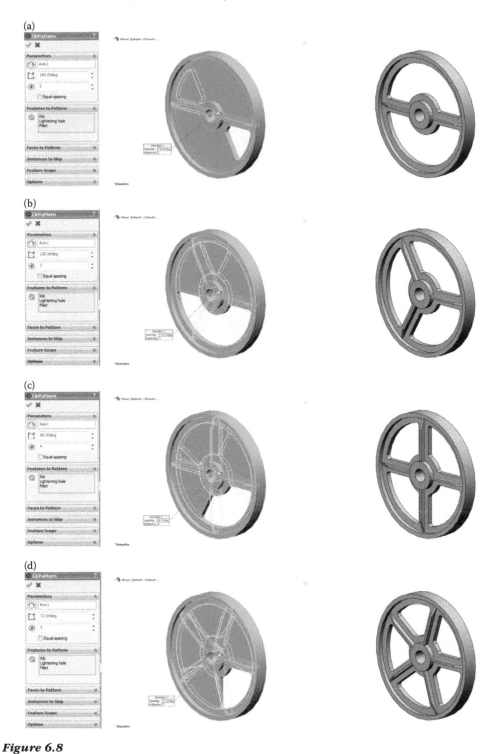

Figure 6.8
Different modified models of the wheel. (a) The number of ribs is 2. (b) The number of ribs is 3. (c) The number of ribs is 4. (d) The number of ribs is 5.

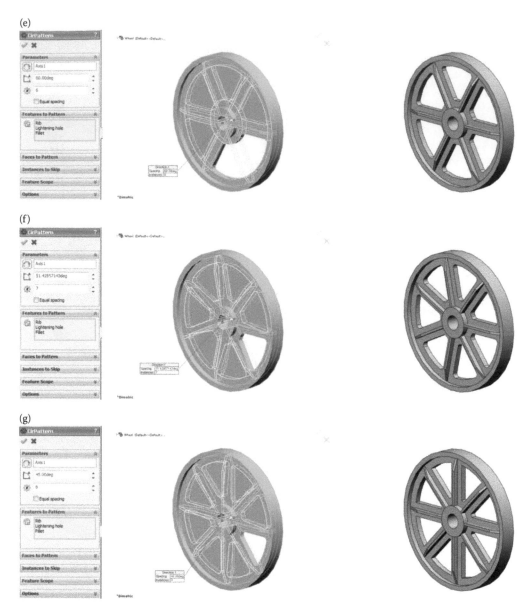

Figure 6.8 (Continued)
Different modified models of the wheel. (e) The number of ribs is 6. (f) The number of ribs is 7.
(g) The number of ribs is 8.

which is set to 60.00 deg for **Sketch5**; Number of instances (✲) – 6; Features to Pattern (⬡): Rib, Lightning hole, Fillet; Faces to Pattern – no selection; Instances to Skip – no selection; Feature Scope: Selected bodies – checked; Bodies to Affect (⬠) – Fillet, Through hole [1], Through hole [3]; Options: Geometry patterns – checked; Propagate visual properties – checked; Partial preview – checked.

The **Graphic area** view and the ready model are shown in Figure 6.7b and c.

16. Defining equations to establish a relation between **Sketch5** and circular patterns (Figure 6.7d):

$$\text{Tools} \rightarrow \text{Equations} \rightarrow \text{OK}$$

Introducing these equations will enable easy modification of the wheel as far as the number of ribs is concerned. To change that number, it is enough to change the data in the **CirPattern** property manager according to Figure 6.8. It is enough to input the **Number of instances** (⚙) and, after that, to modify the **Angle** (⌢) to be equal to **360 deg/Number of instances**.

The **Graphic area** views and views of modified wheels are shown.

It is important to remember that if the number of ribs is 8 or larger, that is, the angle for each instance is 45° or larger. Otherwise, the fillets around the edges of the **Lightning hole** component could not be made due to geometrical interferences (Figure 6.8g).

For further finite element (FE) studies, the number of ribs is chosen to be 6. Its change does not influence the suggested algorithm.

6.1.2 Geometrical Model of a Body with Planar Symmetry

A body, representing a machine unit, which supports two small shafts (Figure 6.9), will be analysed. This machine unit possesses a planar symmetry, which will be used to make the FE simpler and to decrease the computer resources needed for the performance of the analysis.

At first, it will be briefly explained how to develop the CAD model of that machine unit in a SolidWorks environment. The included stages are as follows:

1. Start **a new model**:

$$\text{File} \rightarrow \text{New} \rightarrow \text{Part} \rightarrow \text{OK}$$

Save the file as **Machine_element. sldprt**.

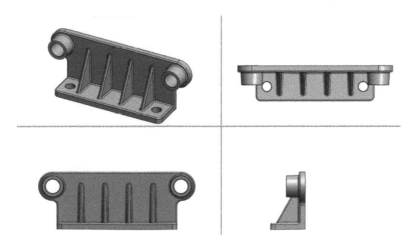

Figure 6.9
CAD model of the analysed machine unit.

2. Define the used units – **System SI** of millimetre, gram and second (MMGS) units is selected.

Tools → Options → Document Properties → Units → MMGS → OK

3. Draw the first sketch in the **Front Plane** (Figure 6.10a):

Sketch → Front Plane → Circle → OK

Automatically the program indexes this sketch as **Sketch1**.
4. Extrude **Sketch1** to feature the **Boss_Extrude_1** component (Figure 6.10):

Features → Extrude Boss/Base → OK

The changed options of the **Extrude Boss/Base** property manager (⬚), which feature that component, are shown in Figure 6.10b. They are From: Sketch Plane; Direction 1: Condition type (⬚) – Blind; Depth (⬚) – 8 mm.
5. Draw the second sketch. The plane of that sketch is the newly created front face of the **Boss_Extrude_1** component (Figure 6.11a and b). The sketch is **a rectangle** with a height of **8 mm** and a width equal to the width of **Boss_**

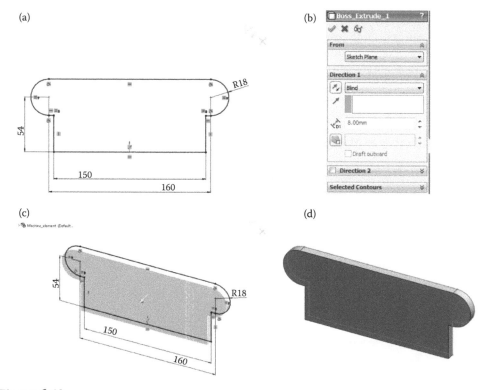

Figure 6.10
Development of the CAD model of the machine element – stages 3 and 4. (a) Sketch1 in Front Plane. (b) Boss_Extrude_1 property manager. (c) Graphic areas view while Extrude property manager is active. (d) Extruded component.

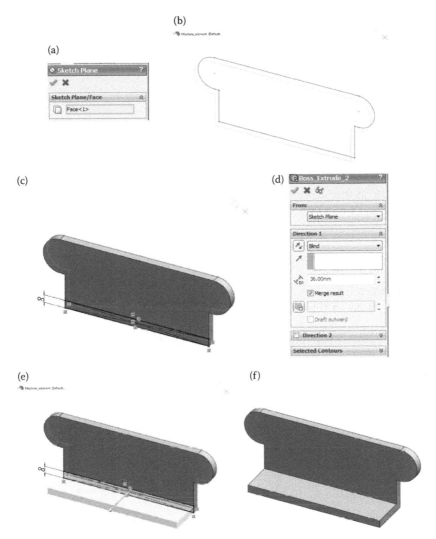

Figure 6.11
Development of the CAD model of the machine unit – stages 5 and 6. (a) Sketch Plane property manager. (b) Picked Face <1>. (c) Ready Sketch2. (d) Boss_Extrude_2 property manager. (e) Graphic area view while Extrude property manager is active. (f) Ready Boss_Extrude_2 component.

Extrude_1. The side lines of the rectangle coincide with the side edges of the created component (Figure 6.11c).

Sketch → Face <1> → Sketch2 → OK

6. Extrude **Sketch2** and establish the horizontal component, denoted **Extrude_ Boss_2** (Figure 6.11d–f):

Features → Extrude Boss/Base → OK

The options of the **Extrude Boss/Base** command (⊞) are shown in Figure 6.11d. They are <u>From:</u> Sketch Plane; <u>Direction 1:</u> Condition type (⊡) – Blind; Depth (⊡) – 36 mm.

7. Feature the fillets at the edges of the components (Figure 6.12a–c):

Features → Fillet → OK

The options of the **Fillet1** property manager (⊡) are <u>Items to Fillet:</u> Constant radius (⊡) – 6 mm; Items to fillet (⊡) – pick the four edges of the model according to Figure 6.12b; Tangent propagation – checked; Full preview – checked; <u>Setback parameters:</u> no data are entered; <u>Fillet Options:</u> Select through faces – checked; Keep features – checked; Overflow type – Default.

8. Sketch the outer contour of the side cylinder (Figure 6.13a). This is a circle with a diameter of 28 mm.

Sketch → Face<1> → Sketch3 → OK

9. Extrude **Sketch3** to make the side cylinder. This component is denoted **Extrude_ Boss_3** (Figure 6.13b–d):

Features → Extrude Boss/Base → OK

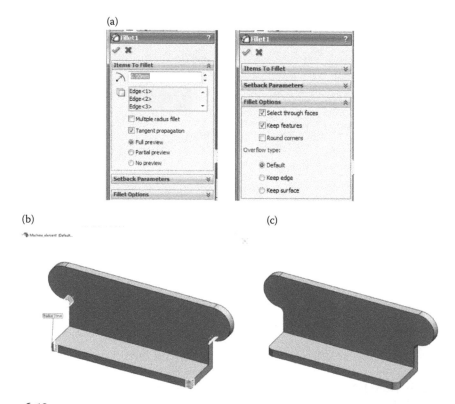

Figure 6.12
Development of the CAD model of the machine unit – stage 7. (a) Fillet property manager. (b) Graphic area view of the picked edges. (c) Filleted components.

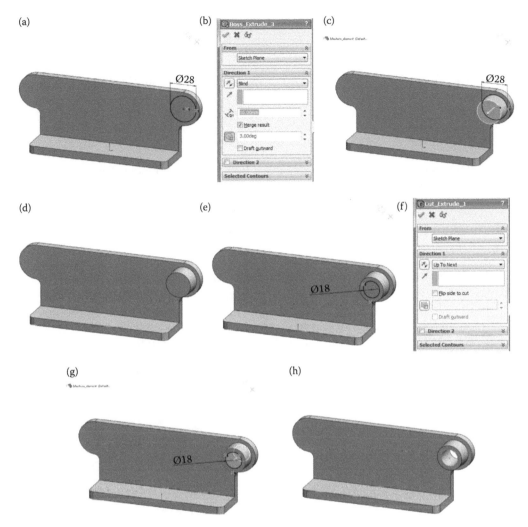

Figure 6.13
Development of the CAD model of the machine unit – stages 8, 9 and 10. (a) Sketching the outer contour of the side cylinder. (b) Boss_Extrude_3 property manager. (c) Graphic area view while Boss_Extrude_3 property manager is active. (d) View of the extruded side cylinder. (e) Sketch of the hole contour. (f) Cut_Extrude_1 property manager. (g) Graphic area view while Cut_Extrude_1 property manager is active. (h) Cut hole inside the side cylinder.

The options of that **Extruded Boss/Base** command (▣) are shown in Figure 6.13b. They are From: Sketch Plane; Direction 1: Condition type (⬝) – Blind; Depth (⬝) – 16 mm; Draft (⬝) – 3.00 deg. This supposes a 3° inward draft to be added to the extruded feature.

10. Cut the hole inside the side cylinder (Figure 6.13e–h). The **Extruded Cut** (▣) command is used to cut the hole through the cylinder and through the component **Boss_Extrude_1**.

Feature → Extruded Cut → OK

The options of the property manager are shown in Figure 6.13f.

11. Feature **Chamfer** (⬛) along the inner front edge of the cylinder (Figure 6.14):

Feature → Chamfer → OK

The chamfer properties, defined using an angle (⬛) and a distance (⬛) options of the property manager, are shown in Figure 6.14a.

12. Sketch the initial point/centre of the side hole:

Sketch → Face<1> → Sketch7 → OK

The sketch is drawn in the bottom face horizontal component (Figure 6.15a). A point with the given coordinates is created (Figure 6.15b and c).

13. Defining a hole, using **Hole Wizard** (⬛, Figure 6.15d–h):

Sketch → Hole Wizard → OK

The **Hole Specification Wizard** has two tabs (Figure 6.15d). The **Type** tab defines the type of the hole, while the **Positions** tab locates the position of the hole on planar or non-planar faces. The point defined in **Sketch8** will be used to locate the hole. As there is no predefined list of styles and we do not intend to keep the data for future projects, Favorite is not selected. Hole Type is defined by clicking on the selected type – Hole (⬛); Standard – ISO; type – Drill sizes. Hole Specifications are as follows: Size – φ12.0; End Condition (⬛) – Up To Next. No any Options defined.

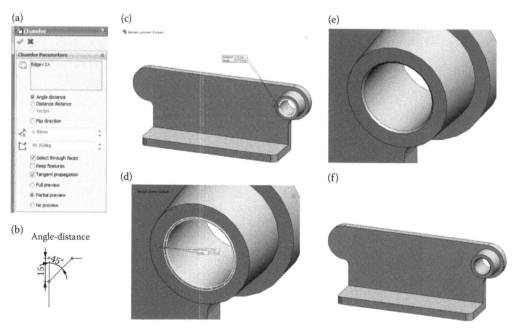

Figure 6.14
Development of the CAD model of the machine unit – stage 11. (a) Chamfer property manager. (b) Angle-distance definition of a chamfer. (c) Graphic area view while Chamfer1 property manager is active, including the introduced chamfer properties. (d) Detailed preview of the chamfered edge. (e) Chamfered edge – detailed view. (f) Chamfered edge.

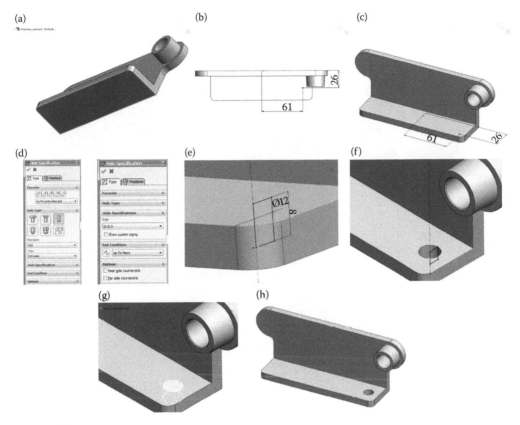

Figure 6.15
Development of the CAD model of the machine unit – stages 12 and 13. (a) Sketching plane for Sketch7. (b) Sketch7 – top view. (c) Sketch7 – trimetric view. (d) Hole Specification wizard. (e) Corresponding to Hole Wizard Sketch8. (f) Cut hole, viewing Sketch8. (g) Graphic area view while Hole Wizard property manager is active. (h) Cut hole.

Sketch8 (Figure 6.15e and f) is generated automatically as the **Hole Wizard** is activated and shows the axis, the diameter and the depth of the hole.

14. Mirror the side cylinder and the hole (Figure 6.16c):

Features → Mirror → OK

The options of the **Mirror1** property manager (▦) are shown in Figure 6.16a. Mirror Face/Plane (▢) is the **Right Plane** and Features to Mirror (▣) are picked according to Figure 6.16b.

15. Define an auxiliary plane, parallel to the **Right Plane** (Figure 6.16d–f):

Features → Reference Geometry → Plane → OK

The options of the **Plane1** property manager (▨) are given in Figure 6.16d. **Plane1** is fully defined.

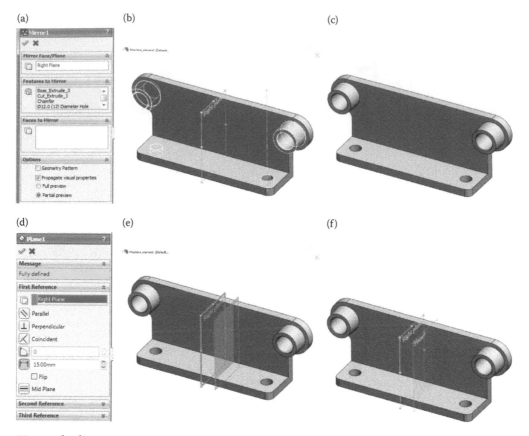

Figure 6.16
Development of the CAD model of the machine unit – stages 14 and 15. (a) Mirror1 property manager. (b) Graphic area view while Mirror1 property manager is active. (c) Mirrored entities. (d) Pane property manager. (e) Plane1 at 15 mm distance of Right Plane. (f) Auxiliary Plane1.

16. Sketch the rib in **Plane1** (Figure 6.17a). This is **Sketch9**. The rib is a triangle with angle of 60° and a side of 60 mm:

Sketch → Plane1 → Sketch9 → OK

17. Feature the rib (🔳) from **Sketch9** (Figure 6.17b). The rib is 4 mm thick and is formed as the sketch is in the middle of the volume (Figure 6.17c and d):

Features → Rib → OK

18. Feature the fillets at the front edges of the ribs (Figure 6.18a–c):

Features → Fillet → OK

The options of the **Fillet2** property manager (🔲) are <u>Items to Fillet</u>: Constant radius (🔗) – 1.5 mm; Items to fillet (🔲) – pick the front edges of the

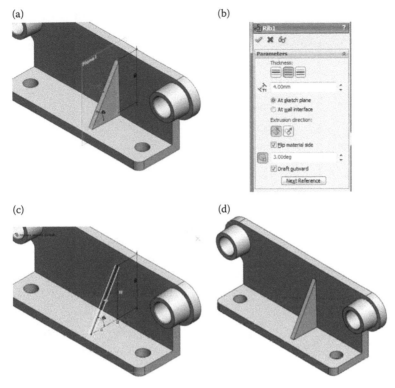

Figure 6.17
Development of the CAD model of the machine unit – stages 16 and 17. (a) Sketch9 in Plane1. (b) Rib1 property manager. (c) Graphic area view while Rib1 property manager is active. (d) Ready Rib1 view.

ribs according to Figure 6.18a; Tangent propagation – checked; Full preview – checked, etc.

19. Pattern the rib in one direction. The **Linear Pattern** (⊞) command is used (Figure 6.19a–c).

Features → Linear Pattern → OK

The options of the **LPattern1** property manager (⊞, Figure 6.19a) are <u>Direction 1:</u> Direction (⟋) – set to be parallel to **Edge<1>**, which is the blue crossing edge of both **Boss_Extruded** components; Spacing distance for patterns (⟋) – 30 mm; Number of Instances (⟋) – 2, including the original pattern; <u>Direction 2:</u> – no definition; <u>Features to Pattern</u> (⟋) – Rib1; Fillet 2; <u>Faces to Pattern</u> (⟋) – no faces to pattern; no <u>Instances to skip</u> (⟋); <u>Options:</u> Propagate Visual Properties and Partial preview are checked.

20. Mirror the patterned ribs (⟋, Figure 6.19d–f):

Features → Mirror → OK

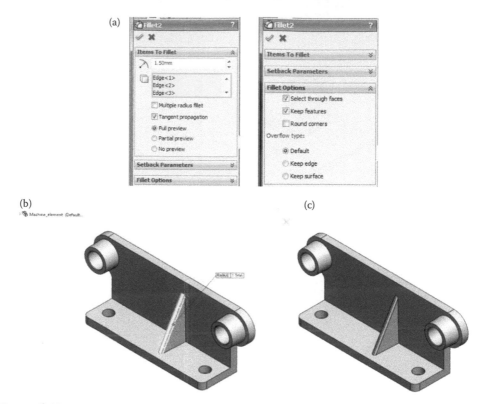

Figure 6.18
Development of the CAD model of the machine unit – stage 18. (a) Fillet2 property manager. (b) Picked edges to be filleted. (c) Filleted front edges of the ribs.

The options of the **Mirror2** property manager are (Figure 6.19d): Mirror Face/Plane (⬜): **Right Plane**; Features to Mirror (⬛): **Lpattern1** in the **Feature Manager** design tree is picked. The same signature is automatically displayed in the blue window; Options: Propagate Visual Properties and Partial preview are checked.

21. Feature the fillets at the edges of the components (Figure 6.20a–c):

Features → Fillet → OK

The options of the **Fillet3** property manager (⬛) are Items to Fillet: Constant radius (⌒) – 1.5 mm; Items to fillet (⬜) – the front vertical and the top horizontal faces of the components Boss_Extruded_1 and Boss_Extruded_2 according to Figure 6.20b; Tangent propagation – checked; Full preview – checked, etc.

Further, more versions of that machine unit are suggested (Figures 6.21 and 6.22). The number and the distance between the ribs are changed through the commands **Plane1** (stage 15) and **LPattern1** (stage 19). The ratio of the distance between **Right Plane** and **Plane1** and the distance between the patterned ribs should always be 1:2. To design the fillets around faces (stage 21) successfully, the

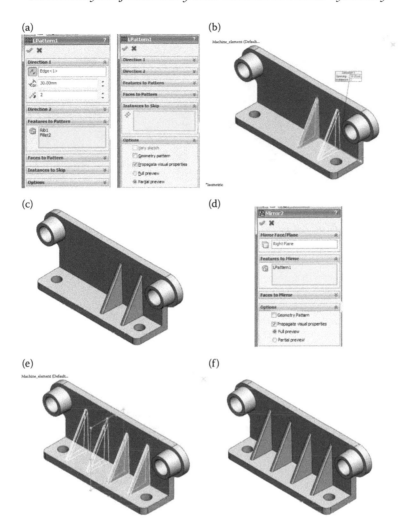

Figure 6.19
Development of the CAD model of the machine unit – stages 19 and 20. (a) LPattern1 property manager. (b) Graphic area view while LPattern1 property manager is active. (c) Patterned ribs. (d) Mirror2 property manager. (e) Graphic area view while Mirror2 property manager is active. (f) View of the mirrored ribs.

distance between the end ribs, the hole and the fillet radius should be carefully calculated. Otherwise, no fillets will be set (Figure 6.22a and b).

We designed two CAD models of machine units in this section. These units have a circular and a planar symmetry.

We suggested a few modifications of these units varying the patterned commands used.

We learned how to modify symmetric CAD objects varying **CirPattern** or **LPattern** commands.

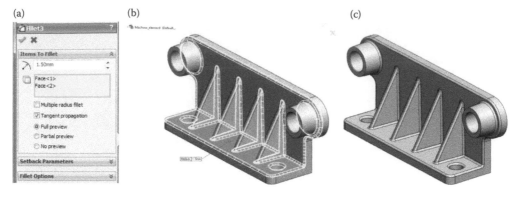

Figure 6.20
Development of the CAD model of the machine unit – stage 21. (a) Fillet3 property manager. (b) Picked faces to be filleted. (c) Filleted components.

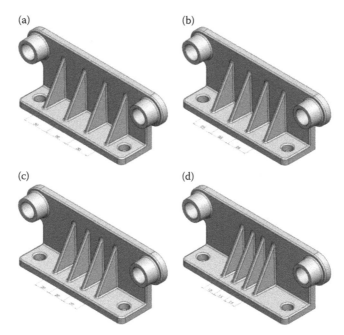

Figure 6.21
Some variation of this machine unit with 4 ribs. (a) Machine unit with 4 equidistant ribs at 30 cm (analysed further detail). (b) Machine unit with 4 equidistant ribs at 25 cm. (c) Machine unit with 4 equidistant ribs at 20 cm. (d) Machine unit with 4 equidistant ribs at 15 cm.

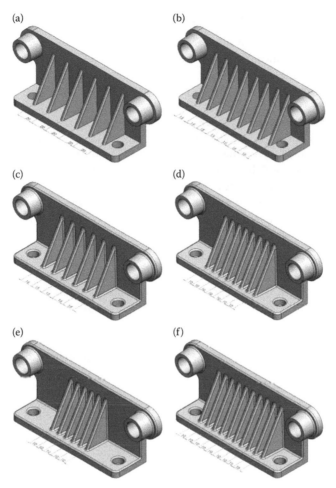

Figure 6.22

Few more variations of this machine unit with 6, 8 or 10 equidistant ribs. (a) Machine unit with 6 equidistant ribs at 20 cm. (b) Machine unit with 8 equidistant ribs at 15 cm. (c) Machine unit with 6 equidistant ribs at 15 cm. (d) Machine unit with 8 equidistant ribs at 10 cm. (e) Machine unit with 6 equidistant ribs at 10 cm. (f) Machine unit with 10 equidistant ribs at 10 cm.

6.2 STATIC ANALYSIS OF THE DESIGNED SYMMETRICAL MACHINE UNIT WITH CIRCULAR SYMMETRY

6.2.1 Why Use Symmetry and How It Works

Symmetry helps us in studying a segment of the model instead of the full model. It requires that geometry, restraints, loads and material properties of the model be symmetrical. The results of the 'missing' segments are deducted from the studied segment by the user, and thus, the 'entire' situation can be analysed. The symmetry helps us to reduce the size of the problem without any decrease in the results' accuracy. The procedures of the application of the symmetrical restraints to solid meshes or to shell meshes using mid-surface are identical.

There are two main groups of symmetrical objects to be discussed:

- **Objects with a planar symmetry** where a segment of the object generated through a few cuts along symmetric planes is studied. Generally, this is a half or a quarter of the entire object. **Symmetry restraints** are applied to the cut sides to guarantee that the face is prevented from moving in its normal direction.

- **Objects with a circular symmetry**, or the so called axi-symmetrical objects. To analyse such a model, a single wedge can be used. Although the angle of the wedge is arbitrary in theory, using a very small angle may result in bad FEs at the tips, especially when there is no hole at the centre of the model. The **symmetrical restraints** are applied to the cut sides of the wedge, that is, to the faces of the symmetry. For solid models, they guarantee that every face that is coincident with a plane of symmetry is prevented from moving in its normal direction.

Objects with a circular symmetry are a part of the larger group of the axi-symmetrical objects. When modelling such an object, circular patterns are used. When designing such an object, a representative segment can be studied. This segment can be a part or an assembly. The geometry, the restraints and the loading conditions are similar for all other segments that make up the model. Turbine, fans, flywheels and motor rotors can usually be analysed using circular symmetry.

The Circular Symmetry restraints can be applied for solid models and for static studies only. To define it, two similar planar sections and the axis of revolution for the symmetry must be defined. The program enforces equal translations at each pair of nodes, which possess similar relative positions on the two sections, that is, nodes on opposite sections with similar relative positions displace similarly. If the loads are such that the cut sections deform normally to their planes, **Circular Symmetry** restraints should be used. If the model has a circular pattern and the loads are such that the cut sections cannot deform normal to their planes, **Symmetry constraints** can be applied on the cut section.

Circular symmetry is more general as it can solve problems where the cut sections can deform in the deform in the tangential and in the normal directions.

For example, any wedge of the disc can be analysed (Figure 6.23a). If all forces are radial, the **Symmetry restraints** can be applied. If tangential loads exist, the cut sections can deform out of their planes and the **Circular Symmetry** must be used. For analysing a fan (Figure 6.23b), just one blade can be analysed. Since loads on the blades are usually tangential, using the **Circular Symmetry** is recommended. Although any one-ninth of the model is a valid pattern, it is recommended to use a

(a)   (b)

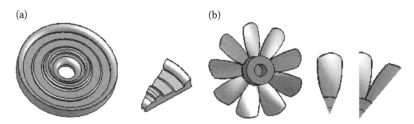

Figure 6.23
Some examples of axi-symmetrical objects (SW Simulation on-line help). (a) A disc and a studied wedge. (b) A fan and two corresponding segments to be studied.

pattern that does not cut through the blades, since the blades are exposed to spatial loading and deformations.

6.2.2 Defining the Analysed Segment

We will now analyse the wheel designed in Section 6.1.1. This is a machine unit with a circular symmetry. If we analyse the entire model, this will require using much more computer resources and time, considering similar boundary conditions and mesh density.

Thus, if we have to analyse an object with a circular symmetry, it is always recommended to try to reduce the model relying on the assumption of existing symmetry.

As the wheel is an object that is axi-symmetric, it is enough to analyse only the one-sixth segment of the model.

At first, we have to identify that segment. The cuts can pass either through the ribs (Figure 6.24a) or through the middle of the lightning holes (Figure 6.24b). The advantages of the first identified segment compared to the second one concern the deformations and the stress–strain distribution inside the wheel in relation to the loading in the provided additional examples. As the cuts of the segment pass through the ribs, that is, through the more rigid components of the object, the expected boundary deformations will be less. Thus, the impact of the definition of symmetrical boundary conditions and of the analysis of a segment of the wheel instead of the entire wheel will be reduced compared to the version when cuts pass through lightning holes (Figure 6.24b).

The next stage is cutting the identified segment. To do so, further steps should be performed:

1. Defining the intersection of the wheel with the **Top Plane** (Figure 6.25a and b)

Sketch → Convert Entities (⬚) → Intersection Curve (▨) → OK

Pick the **Top Plane** from the floating **Design Tree** at the **Graphics area** as well as all faces that the **Top Plane** intersects (Figure 6.25d). Pick the faces of the rib that lies at 180° of the initial one. Otherwise, you could have problems; creating the mesh for the software automatically generates two interfering bodies. Their signatures are automatically displayed in the blue window in the **Intersection Curve** property manager (Figure 6.25c). After clicking **OK**,

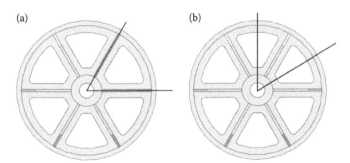

Figure 6.24
Identifying the analysed segment. (a) Segment with cuts through the ribs. (b) Segment with cuts through the lightning holes.

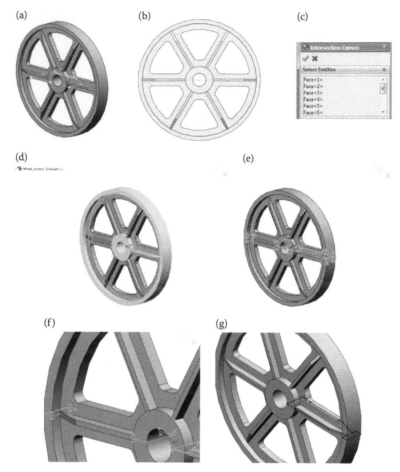

Figure 6.25
Defining the intersecting contour – Sketch6. (a) Trimetric view of the intersection of the wheel and Top Plane. (b) Front view of the intersection of the wheel and Top Plane. (c) Intersection Curves property manager. (d) All picked intersected faces. (e) Intersection curve, defined as an open sketch. (f) Intersection curve, defined as an open sketch – detailed view. (g) Sketch6.

the intersecting curve appears as an open contour (Figure 6.25e). The lines of that contour that are at the opposite rib and remain open must be deleted as a prerequisite to the success of the next operation (Figure 6.25f).

Sketch → Trim (⌗) → Trim to closest (⊢) → OK

The **Trim** command is used to draw the sketch as a closed contour (**Sketch6**, Figure 6.25g).
2. Featuring the segment with the **Cut-Revolve** (⌾) command

Feature → Cut-Revolve → OK

The options of the **Cut-Revolve** property manager are given in Figure 6.26a. The section will be revolved around **Axis1** (<u>Axis of Revolution,</u> ⬉), clockwise

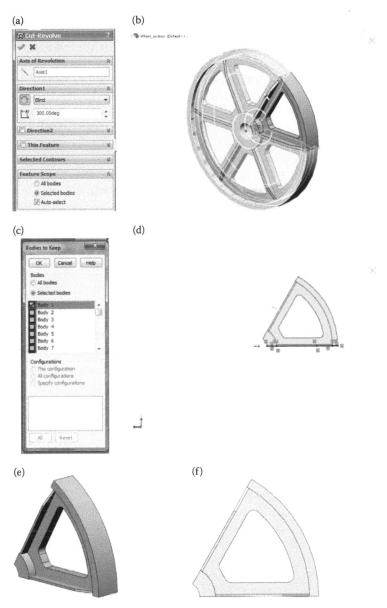

Figure 6.26
Cutting the segment to be studied. (a) Cut-Revolve property manager. (b) Graphic area view while Cut-Revolve property manager is active. (c) Bodies to keep property manager. (d) Selected body to be kept – coloured in blue. (e) The cut segment – trimetric view. (f) The cut segment – front view.

(Blind, ⊙), at an angle (⬚) of 300deg. The cut material is coloured in yellow in the **Graphics area** and the kept bodies – in grey (Figure 6.26b). Clicking on the **OK** button opens a new **Bodies to Keep** window (Figure 6.26c), where all bodies are listed, and the user must check the signatures of the bodies to be kept. Only the first and the second bodies, which are coloured in blue in Figure 6.26d, will be kept, that is, only their signatures are selected.

Views of the cut segment are shown in Figure 6.26e and f.

6.2.3 Static Study of a Body with Circular Symmetry and Symmetrical Loads

The segment will be studied under symmetrical loading. If the wheel operates as a transferring element between a pipe and a shaft, in which the diameter is much smaller than the inner diameter of the pipe (Figure 6.27a), such a type of loading will be generated. It is supposed that the wheel is steadily pressed in the pipe. Thus, the inner surface of the pipe will generate **a uniform pressure** normal to the outer surface of the wheel. The thin shaft acts as a supporting component of the entire wheel–pipe unit. Hence, it can be modelled as **a fixture**, whose type will be specified further in this study. The free-body spatial scheme of the wheel, including the loading (the red arrows) and the fixture (the green symbols), is shown in Figure 6.27b.

To perform the static study of the wheel, only one-sixth segment will be used. It has already been shown how this segment could be cut from the entire model.

A new study titled **Symmetrical_Study_Fixed_Geometry** is started.

The development of the **FE model** starts with setting the material. It is the **Gray Cast Iron**, selected from the **SW Materials** library:

Name of the part (right click) → Apply/Edit Material (⌘☰) → SolidWorks Materials → Iron → Gray Cast Iron → Apply → Close

The **Gray Cast Iron** is a linear isotropic material, which have the following material properties: modulus of elasticity – 66,178 MPa; Poisson's ratio – 0.27; shear modulus – 50,000 MPa; mass density – 7200 kg/m³; tensile strength – 151.66 MPa; compressive strength – 572.17 MPa. The cast iron is **a brittle material**, and hence no yield strength is defined and the default failure criterion is according to the **Mohr–Coulomb theory**.

Further, the pressure loads are applied:

SW Simulation analysis tree → External Loads (right click) → Pressure (⊞)

A uniform pressure, normal to the selected surface, will be applied to the outer surface of the wheel (Figure 6.28b) with a value of 5 MPa (Figure 6.28a).

The next task is to define the fixtures. At least two types of fixtures should be defined. The first type replaces the shaft impact on the wheel, and the second type 'tells' the software that the studied segment is a piece of a larger model. It is questionable how exactly the shaft interacts with the wheel and which is the best fixture to be

(a)   (b)

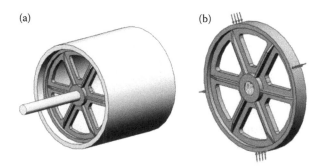

Figure 6.27
Studied shaft wheel–pipe unit. (a) CAD model of the machine unit. (b) Free-body scheme of the wheel.

(a)   (b)

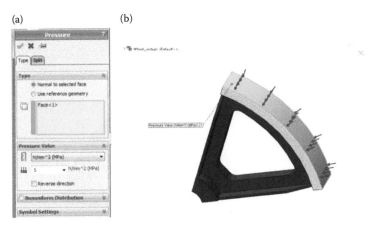

Figure 6.28
Applying the pressure. (a) Pressure property manager. (b) Pressure symbols and loaded face.

applied. This strongly depends on the type of joining of both elements. Thus, three different versions of the union are modelled and studied:

- If we assume that the wheel is fixed to the shaft and the shaft is **rigid**, the shaft can be replaced with the **Fixed Geometry** fixture. This study is titled **Symmetrical_Study_Fixed_Geometry**. The effect is similar if a rigid **Bearing** fixture is introduced.
- Another option to be studied is as follows: if we assume that the joining of the wheel to the shaft is implemented through a damper with known damping properties, and enable a free radial, circumferential or axial motion within predefined limits, then **Advanced Fixture on Cylindrical Faces** is used. This study is titled **Symmetrical_Study_Advanced_Fixture**.
- If we assume that the shaft is deformable and there is a bearing connector between the shaft and the wheel, a **Bearing Fixture** is recommended. Probably, this is the most common case if torsion is applied. This study is titled **Symmetrical_Study_Bearing_Fixture**.

The first study to be performed is the **Symmetrical_Study_Fixed_Geometry** study, where a **Fixed Geometry** fixture is applied (Figure 6.29):

SW Simulation analysis tree → Fixtures (🔲, right click) → Fixed Geometry (🔏)

While the first fixture is different for all case studies, the second fixture replaces the 'missing' part of the wheel and is similar in all studied cases. This is a **Circular Symmetry** fixture, which belongs to the group of **Advanced Fixtures** (Figure 6.30a):

SW Simulation analysis tree → Fixtures (🔲, right click) → Advanced Fixtures → Circular Symmetry (🔘)

The faces to which the restraints are applied are the side faces of the segment and are coloured in blue and in violet for each cutting plane (Figure 6.30b). The symmetrical axis is introduced by selecting **Axis1** from the floating **Design tree** on the **Graphics area** and is coloured in pink (Figure 6.30).

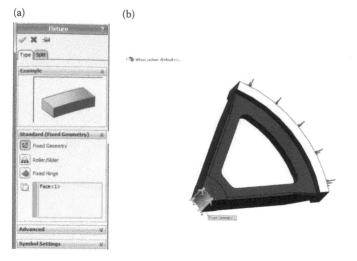

Figure 6.29
Applying the Fixed Geometry fixture to the wheel segment. (a) Fixture property manager – Fixed Geometry fixture. (b) Fixed Geometry face.

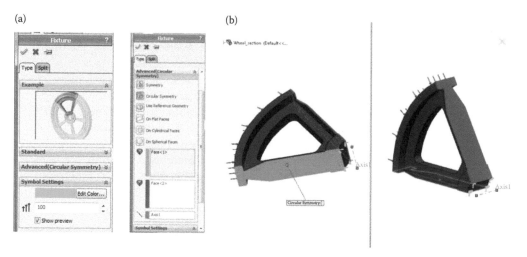

Figure 6.30
Applying the Circular Symmetry fixture. (a) Circular Symmetry fixture property manager. (b) Faces, where Circular Symmetry boundary conditions are applied and the corresponding circular axis.

Another alternative to define the symmetry restraints in this case is the use of the **Symmetry** fixture (Figure 6.31a):

SW Simulation analysis tree → Fixtures (⬛, right click) → Advanced Fixtures → Symmetry (⬛)

The faces to which these new restraints should be applied are the side faces and are coloured in blue (Figure 6.31b). This alternative is applicable only if the applied loading does not deform the symmetrical faces out of their planes.

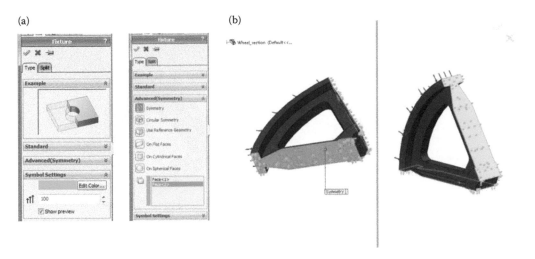

Figure 6.31

Applying the Symmetry fixture. (a) Symmetry fixture property manager. (b) Faces, where Symmetry boundary conditions are applied.

Yet, considering our intention to use the developed FE model for further analysis where the wheel will be exposed to torsion **'Circular Symmetry,'** a fixture is preferred.

The next stage is to define mesh properties. Generally, this operation is done after fixtures and loads are defined, as they are a part of the defined boundary conditions. The mesh that will be created at that stage will be used within all the following studies. The options of the **Mesh** property manager are shown in Figure 6.32a. A curvature-based mesh with the following properties is generated (Figure 6.32b): Jacobian points – 29; Max Element Size – 6 mm; Min Element Size – 2 mm; Total nodes – 19,899; Total elements – 11,836 (Figure 6.32c). The highest Aspect ratio is 530.8, and it is related to a few elements at the joint between the rib and the rim (the red areas, Figure 6.32d).

If you have any problems with mesh generation, this could be due to the interference or the existence of more than one component, which the software generates automatically within the model. The easiest way to overcome that problem is to create incompatible meshes. This can be done in a few different ways:

- Through the **Component Contact** property manager – Start **Component Contact** property manager (Figure 6.33a).

SW Simulation analysis tree → Connections (⬚⬚, right click) → Component Contact (🖳) → OK

Then select the **Bonded (No clearance) Contact type** to ensure the continuity of the model and the transfer of loads between the selected bodies, that is, it guarantees that the selected bodies will behave as if they were welded during simulation. The program creates a **compatible mesh** on contacting areas. This means that the program merges coincident nodes along the common interface (Figure 6.33c). Bonding with a compatible mesh produces more accurate results in the bonded regions than bonding with an incompatible mesh, but it can cause meshing to fail for some assemblies. If so, the remeshing of the failed parts with an incompatible mesh can help (Figure 6.33d).

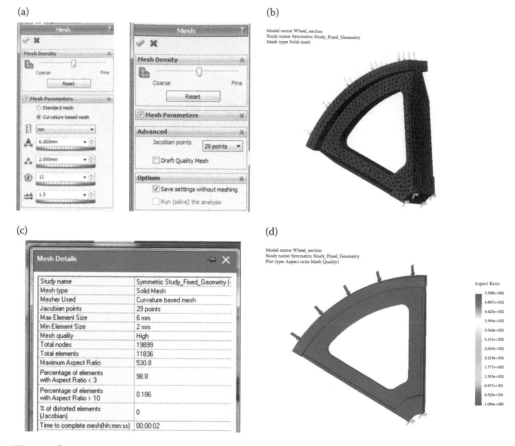

Figure 6.32
Meshing the object. (a) Mesh property manager – curvature-based mesh. (b) Plot of the created mesh. (c) Mesh details. (d) Aspect Ratio plot.

Then, the program meshes each component independently and uses multi-point restraints internally. Bonding incompatible meshes can generate local stress concentrations in the bonded areas. The **Bonding incompatible mesh** is activated for every component contact separately.

- Through the **Mesh** property manager – Check **Remesh failed parts with incompatible mesh** in the **Advanced** sub-window (Figure 6.33b).

SW Simulation analysis tree → Mesh (right click) → Create Mesh (🖻) → OK

- Through the **Simulation toolbar**, following the path (Figure 6.33c)

Simulation → Option → Default Options → Mesh →Mesher Settings → Remeshed failed parts with incompatible mesh

If this option is on, the options are active for the entire part/assembly and the software tries incompatible meshing automatically for solids that fail to mesh with the compatible option.

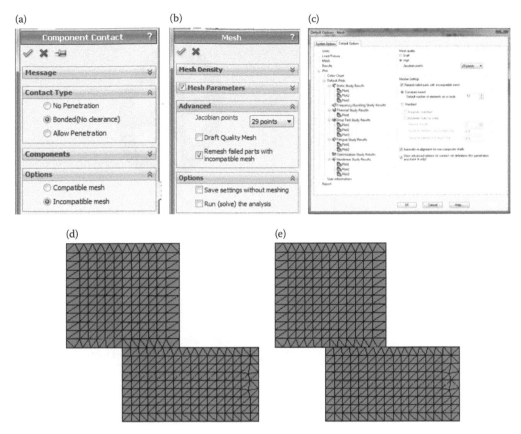

Figure 6.33
Bonding bodies (SW Simulation On-line Help). (a) Component Contact property manager. (b) Mesh property manager. (c) Mesh options. (d) Bonding with compatible mesh. (e) Bonding with incompatible mesh.

Finally, the object is meshed and the study is run. The FE model has 59,262 DOFs and the calculation runs for about 9 s. All the above model properties are the same as the properties of the FE model of the entire wheel, are exposed to the same restraints and are meshed with a similar mesh (Figure 6.34).

The wheel mesh consists of 105,500 (compared to 19,899 for the wheel segment) nodes and 65,376 (compared to 11,836) elements. The highest **Aspect Ratio** is 853.68 (compared to 530.8) and is for the few elements at the joints between the ribs and the rim as is the wheel segment model. The solver runs a model of 314,490 (compared to 59,262) DOFs for about 12 s (9 s runs the wheel segment model). All values in the brackets are for the segmented wheel model. Now, it is easy to compare the complexity of both models, especially regarding the necessary computer resources. As the FE model is a relatively easy model, its complexity is not a crucial point for the study.

However, if a more complicated assembly model is studied, the use of circular symmetry to simplify the FE model can be of real help.

Table 6.1 provides the results of three different case studies concerning the complexity of the FE models – the entire wheel model and the two wheel segment models. The percentage of discrepancy is calculated through comparing the values of segmented models to those for the entire wheel. All provided values show a good coincidence.

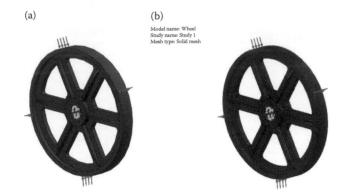

Figure 6.34
FE model of the entire wheel. (a) Model restraint. (b) Plot of the mesh.

Table 6.1
Numerical Results for the Entire Wheel and for the Segmented Wheel Models

	Entire Wheel Model	Segment, Formed by Cuts through the Ribs		Segment, Formed by Cuts through Lightening Holes	
	Value	Value	Discrepancy	Value	Discrepancy
Von Mises Stress, Max Element Mode (MPa)	94.3	92.9	1.48	94.1	0.21
Von Mises Stress, Max Node Mode (MPa)	110.3	106.3	3.69	105.9	3.99
Principal Stress P1, Max Element Mode (MPa)	11.2	11.1	−0.89	11.2	0
Principal Stress P3, Min Element Mode (MPa)	−98.7	−98.1	0.61	−99.1	−0.41
Maximum Displacement (mm)	8.751e−02	8.75e−02	0	8.753e−02	−0.02
Factor of Safety, Minimal Value	5.1	4.04		5.1	

The extreme values of P1 and P3 in the node mode appear in one of the nodes of the FE with the highest aspect ratio. Therefore, these results are not reliable and they are omitted in the table.

Finally, we can conclude that the numerical calculations for segmented models can successfully replace those for the entire complex model. However, it is necessary to keep in mind that the user himself or herself has to transfer the segmented results to the entire model.

The corresponding plots are shown in Figure 6.35.

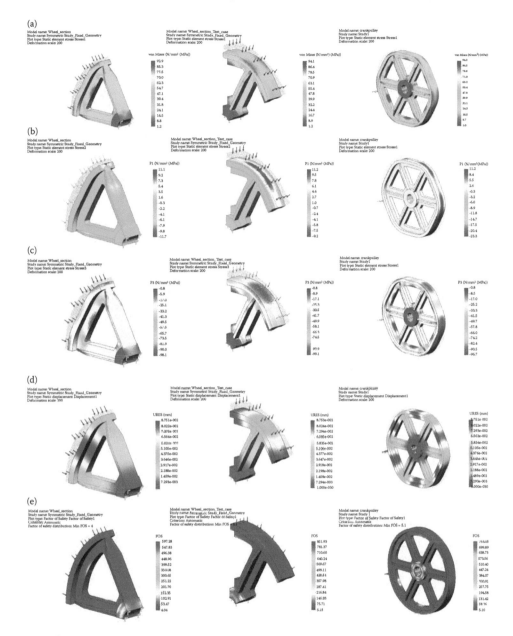

Figure 6.35
Plots of the results of the three models. (a) Plot of von Mises stresses. (b) Plot of principle stresses P1.
(c) Plot of principle stresses P3. (d) Plot of the displacement. (e) Factor of safety plots.

Further, the results of some additional studies with different fixtures to model the supporting shaft are given. The studied models are as follows:

- **Study 1** – It is titled **Symmetrical_Study_Advanced_Fixture** and an **Advanced Fixture on Cylindrical Faces** is applied. The properties of the fixture are given in Figure 6.36a. It enables radial motion of 0.02 mm and torsion of 0.5° (Figure 6.36b). Some plots of the result are also shown (Figure 6.36c–h).

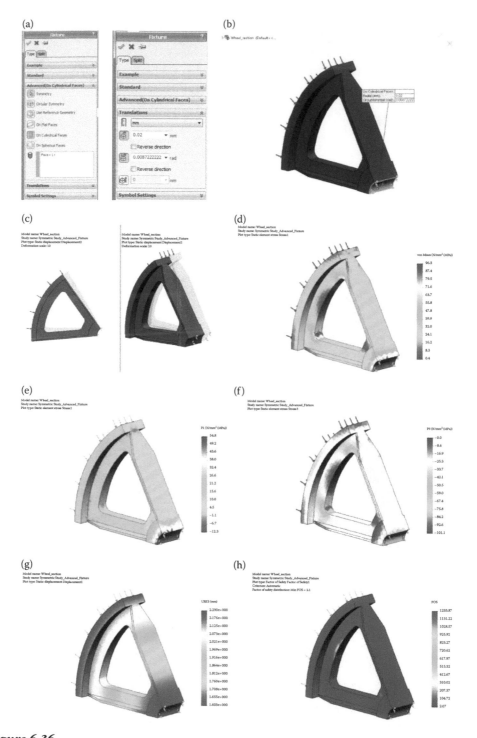

Figure 6.36
Plots of the results of Symmetrical_Study_Advanced_Fixture. (a) On Cylindrical Faces property manager. (b) Graphical view of On Cylindrical Face fixture. (c) Deformed shape. (d) von Mises plot. (e) First Principle Stress plot. (f) Third Principle Stress plot. (g) Displacement plot. (h) Factor of Safety plot.

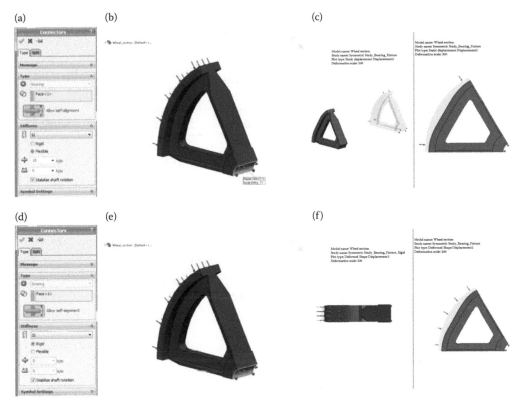

Figure 6.37
Applying different bearing fixtures. (a) Bearing Fixture property manager. (b) Graphic view of the flexible bearing. (c) Deformed shape of flexible Bearing_Fixture study. (d) Bearing Fixture property manager – rigid bearing. (e) Graphic view of the rigid bearing. (f) Deformed shape of rigid Bearing_Fixture study.

- **Study 2** – This study is titled **Symmetrical_Study_Bearing_Fixture** and uses the **Bearing Fixture** (⚙, Figure 6.37a and b). The **Self-aligning** option (⇌), which allows an unrestricted off-axis shaft rotation, is active. The fixture is **Flexible** with **Radial** resistance (⇕) of 10 N/m and no ability to resist to **Axial** displacement (⇔). To prevent rotational instability, caused by torsion, and to avoid numerical singularities, the **Stabilize shaft rotation** is checked. The deformed form is shown in Figure 6.37c. The fixture allows free motion along the axis of the wheel. Regarding that criterion, this fixture can be compared to the **Cylindrical Faces** fixture, which also frees that motion in the input limits. The difference is in the way of introduction of these limits – either through the displacement range for the **Cylindrical Faces** fixture or through axial and radial resistance for the **Bearing** fixture.
- **Study 3** – It is titled **Symmetrical_Study_Bearing_Fixture_Rigid** and the applied **Bearing Fixture** (⚙, Figure 6.37d and e) is **Rigid**. The deformed form is shown in Figure 6.37f. It is obvious that regarding the deformations, the rigid bearing acts as a **Fixed Geometry** fixture.

Table 6.2

Numerical Results for the Segmented Models, Restrained by Different Fixtures (Scenario 1)

	Study with Fixed Geometry Fixture	Study with Cylindrical Faces Fixture, Study 1	Bearing Fixture Studies with Axial Bearing Resistance Equal to		
			10 N/m Study 2	5 N/m	0 N/m Study 3
von Mises Stress (MPa)					
Element mode – max	92.9	95.3	92.9	92.9	92.9
Node mode – max	106.3	108.5	106.3	106.3	106.3
Principal Stress P1 (MPa)					
Element mode – max	11.1	54.8	11.1	11.1	11.1
Node mode – max	33.7	61.1	34.4	34.4	34.4
Principal Stress P3 (MPa)					
Element mode – min	−98.1	−101.2	−98.1	−98.1	−98.1
Node mode – min	−111.4	−114.0	−111.4	−111.4	−111.4
Maximum displacement (mm)	8.751e−02	2.23	1.678	1.615	8.751e−02
Corresponding axial displacement (mm)	4.860e−03	1.600	1.676	1.613	4.860e−03
Min Factor of Safety	4.04	2.07	3.97	3.97	3.98

The extreme values of the stress, the displacement and the minimum factor of safety of all these studies are given in Table 6.2. It must be kept in mind that despite their different flexible properties, bearing fixtures produce equal stress distribution.

6.2.4 Static Study of a Body with Circular Symmetry and Anti-Symmetrical Loads

When the pipe is exposed to pure (only to) torsion, the wheel in Figure 6.27a must be studied under anti-symmetrical loading.

Regarding the existence of a steady pressed joint between the pipe and the wheel, the **uniform pressure** normal to the outer surface of the wheel is preserved. Additionally, a **torsional moment** substitutes the torsion to which the pipe is exposed and which is transferred to the wheel. Once more, a wedge of the wheel will be studied.

In addition to the previously introduced pressure, a uniform torque (⊞) of 500 N m is introduced (Figure 6.38).

SW Simulation analysis tree → External Loads → Torque

The results of the five compared studies are provided in Table 6.3 and in Figure 6.39. The impact of applying different fixtures is obvious.

The fixture **On Cylindrical Faces** influences stresses and deformations, while the bearing fixtures generate an entirely different effect.

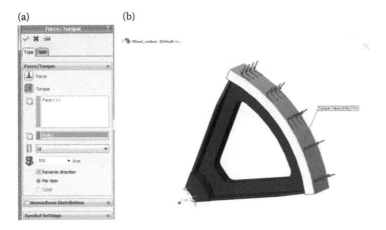

Figure 6.38
Input of the torque. (a) Torque property manager. (b) Graphical view when Torque property manager is active.

Table 6.3
Numerical Results for Segmented Models, Restrained by Different Fixtures (Scenario 2)

	Study with Fixed Geometry Fixture	Study with Cylindrical Faces Fixture Study 1	Bearing Fixture Studies with Axial Bearing Resistance Equal to		
			10 N/m Study 2	5 N/m	0 N/m Study 3
von Mises Stress (MPa)					
Element Mode – max	140.5	216.8	140.5	140.5	140.5
Node Mode – max	160.6	236.9	160.6	160.6	160.6
Principal Stress P1 (MPa)					
Element Mode – max	76.5	144.1	76.5	76.5	76.5
Node Mode – max	89.4	159.8	89.4	89.4	89.4
Principal Stress P3 (MPa)					
Element Mode – min	−149.0	−156.4	−148.9	−148.9	−149.0
Node Mode – min	−167.8	−177.8	−167.8	−167.8	−167.8
Maximum Displacement (mm)	0.452	2.353	0.609	0.599	0.452
Corresponding Axial Displacement (mm)	4.971e−03	1.248	0.4053	0.3902	4.971e−03
Min Factor of Safety	1.70	0.81	1.70	1.70	1.70

As far as the stress distribution is concerned, the use of the bearing fixture is equal to the **Fixed Geometry** fixture. Generally, the bearing fixture provides different nodes' displacements, and consequently different deformations of the body. The displacement values depend on the flexibility of the fixture. Therefore, **the rigid bearing** can successfully substitute the **Fixed Geometry**, as far as the object deformations are concerned.

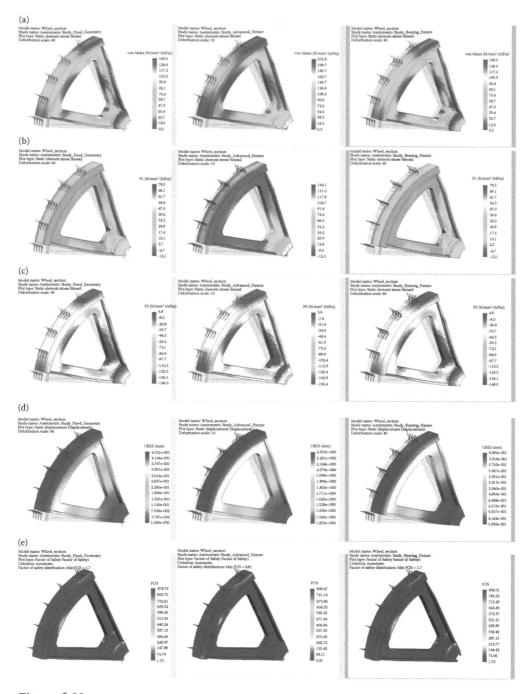

Figure 6.39

Plots of the bearing studies. (a) von Mises Stresses plot. (b) First Principal Stress P1 plot. (c) Third Principal Stress P3 plot. (d) Resultant displacement plot. (e) Factor of safety plot.

In this section, we studied a machine unit with axi-symmetrical geometry, loaded with symmetrical or anti-symmetrical loads, and compared different results. Some test studies were provided.

We learned how to perform a finite element analysis of a segment of a unit with circular symmetry using the symmetry to simplify the model, without any reduction of the results' accuracy. We applied

- Symmetric boundary conditions using Symmetric and Circular Symmetry fixtures
- Bearing fixtures

6.3 STATIC ANALYSIS OF THE DESIGNED SYMMETRICAL MACHINE UNITS WITH A PLANAR SYMMETRY

6.3.1 Defining the Analysed Segment

We start our case study with cutting the half of the body. The new part is titled **Machine_element_section**.

To cut the half of the unit, the next stages are performed:

1. Sketching the intersection of the body with the **Right Plane** of symmetry (Figure 6.40a)

   Sketch → Convert Entities tool (⬚) → Intersection Curve (⊛) → OK

   Pick all intersecting **Right Plane** faces (Figure 6.40b). As a result, their signatures will be displayed at the blue window on the **Intersection Curves** property manager (Figure 6.40c). The software draws the intersecting contour and titles it **Sketch10** (Figure 6.40d).

2. Cutting the unnecessary part. The generated **Sketch10** will be used for extrusion.

   Feature → Extruded Cut (⬚) → OK

   The extrusion will pass through all bodies (Figure 6.41a and b), and only the uncut part of the object (blue body in the figure, Figure 6.41c and d) will be kept for future analysis (Figure 6.41e).

6.3.2 Static Study of a Body with Planar Symmetry and Symmetrical Loads

Let us suppose that there are two small shafts (the orange components) pressed in the horizontal side cylinder components of the units. They are exposed to torsion in two opposite directions. The unit is fixed to the ground by two pins (the blue components in Figure 6.42). Further **Grounded Bolt** fixtures will substitute these pins.

The new static study is titled the **Symmetrical Study**.

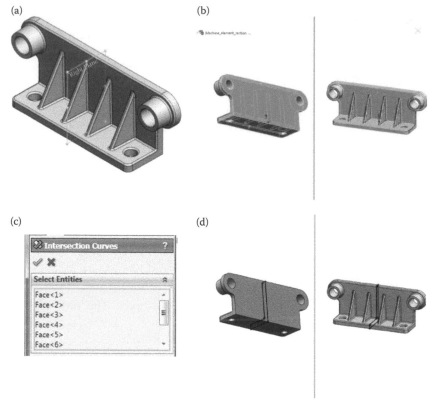

Figure 6.40
*Development of intersection of the object and the symmetrical plane. (a) Plane of symmetry.
(b) Intersected faces. (c) Intersection Curves property manager. (d) Intersected sketch – Sketch10.*

The selected material is the **Gray Cast Iron**. As it possesses very good damping properties, the unit will successfully damp the vibrations of the shafts and will limit the vibrations' impact on the external environment. The material is picked from

SW Simulation Materials → Iron → Gray Cast Iron

Its **Tensile Strength** is 151.66 MPa, the **Compressive Strength** is 572.15 MPa and the **Default failure criterion** is the **Mohr–Coulomb Stress**.

The first fixture to be defined is the **Grounded Bolt** (⊥):

Fixtures → Grounded Bolt…

It sets a bolt connector between the selected component and the ground.

In advance, the **Rigid Virtual Wall** contact must be defined. It is situated in a plane at the bottom of the horizontal component of the unit (Figure 6.43b). Its signature will be **Plane2** (Figure 6.43a and b) and it is created through the path

Feature → Reference Geometry → Plane (◈) → OK

Virtual Wall is created by the following path:

(a) (b) (c)

(d) (e)

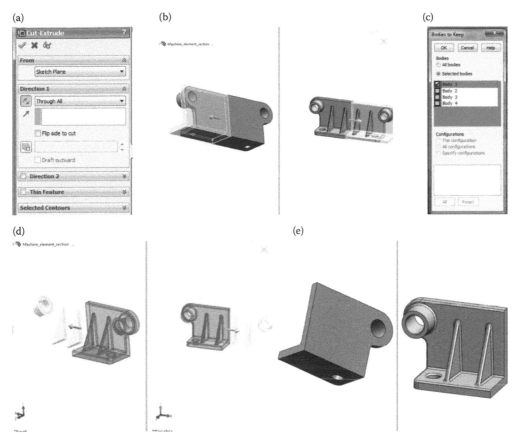

Figure 6.41
Cut half of the object. (a) Cut-Extrude property manager. (b) Graphic area view when Cut-Extrude command is active. (c) Bodies to Keep window. (d) Selected body to be kept. (e) Cut object.

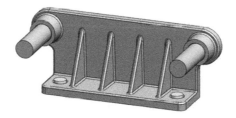

Figure 6.42
CAD model of the studied machine unit.

Connections (🔧) → Contact Sets (🖨) → OK

When the **Contact Sets** property manager opens (Figure 6.43c), the way of introducing the contact entities and the type of the contact (**Virtual Wall** in the **Type** sub-window) must be selected (**Manually select contact sets** in our case). **Virtual Wall** is used to define the contact between the **Set 1** entities (in the blue window of the **Contact Sets** property manager) and the **target plane** (in the violet window). The

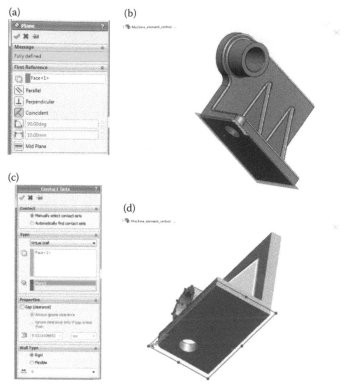

Figure 6.43
Defining the Rigid Virtual wall. (a) Plane property manager. (b) Selected face, which coincides with the plane. (c) Contact Sets property manager. (d) Graphic area view while Virtual Wall is being defined.

target plane may be rigid or flexible. Additionally, a friction between both sets can be defined by entering a non-zero value of the friction coefficient (⚏, Figure 6.43c and d).

The **Ground Bolt** definition may be done according to the following instructions:

1. Setting **the type of the connector** (Figure 6.44a and b). Pick the type of the bolt – Foundation Bolt (🔩); select an edge to define the bolt head and the bolt nut location (◎) – Edge<1>; select a plane to model the virtual wall (◈) – Plane2. The rigid virtual wall prevents the bolt's penetration into the foundation. After that, the values of the **Nut Diameter** (🔩) and the **Bolt Shank Diameter** (🔩) must be entered. The value for the **Bolt Shank Diameter** (🔩) should be equal to or less than the diameters of the **Thread face** – 12 mm. By default, the program multiplies the shank diameter by a factor of 1.5 to obtain the nut diameter – 18 mm.

2. Definition of **Tight Face** (Figure 6.44c) – It is selected if the radius of the shank is equal to the radius of the cylindrical face associated with at least one of the connected components (🔩). The software assumes that the preselected cylindrical face is rigid and that it deforms with the shank as a rigid body. More than one cylindrical face can be selected, but they must have the same axis and radius. This option is not active in our study.

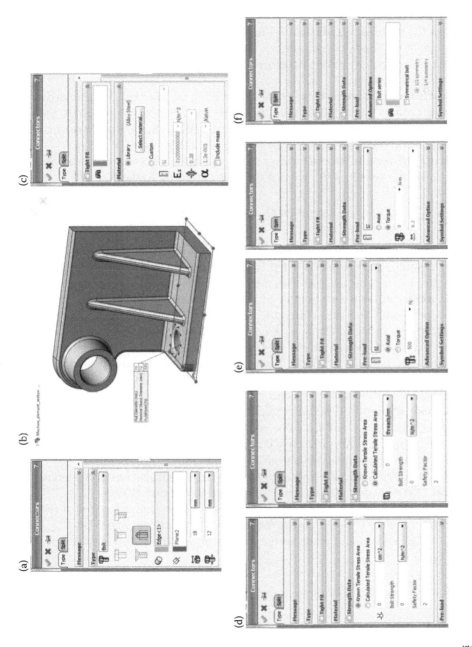

Figure 6.44

Defining the Ground Bolt fixture. (a) Connectors property manager – Type sub-window. (b) Selected the rigid virtual wall, which coincides with Plane2. (c) Connectors property manager – Material sub-window. (d) Connectors property manager – Strength Data sub-window. (e) Connectors property manager – Pre-load sub-window. (f) Connectors property manager – Advanced Option sub-window.

3. Definition of the **Material** (Figure 6.44c) – By default, the material of the bolt is set to be **Alloy steel** from the **SW Simulation Material** library. It can easily be changed by pressing the **Select Material...** (Select material...) button and selecting another material from the library (the **Library** button is checked) or by checking the **Custom** button and defining the properties of a new material. The material properties expected by the program are **Units** (E), **Young's Modulus** (E_x), **Poisson's Ratio** (⚘) and **Thermal expansion coefficient** (α). Optionally the mass of the bolt can be included in the analysis (⚒).

4. The next sub-window is the **Strength Data** (Figure 6.44d) sub-window, which has two versions:
   - **Known tensile stress area** – this option must be selected if **the tensile stress area** is known. The user inputs the area (✄), which equals the minimum area of the threaded section of the bolt (in mm²), **Bolt Strength** (in MPa) and **Safety Factor** for pass/no pass design check of the bolt.
   - **Calculated tensile stress area** – if this option is selected, the program calculates the tensile stress area of the bolt according to the formula

$$A_T = 0.7854*(D_n - 0.9382/n)^2$$

   where A_T equals the tensile stress area, D_n is the nominal shank diameter, p is the thread pitch and $n = 1/p$ is the **Thread Count** (🔩) or threads per millimetre (thread/mm) measured along the length of the fastener. **Bolt Strength** sets the strength of the bolt's material and its unit. There are three commonly used strength parameters for bolts to estimate bolt failure: Yield strength, Ultimate Strength and Proof Strength (90% of Yield strength). The most commonly used parameter of all is the Yield strength of the bolt's material or grade, but the user can choose the most appropriate for the application value. Finally, the **Safety Factor** is input. The bolt fails when its combined load exceeds the ratio of **1/Safety Factor**. The **design check** of the bolt checks whether a bolt can safely carry the applied loads, or it will fail. The software calculates the combined load ratio a connector withstands and compares it with the user-defined factor of safety. All used equations are given in Table 6.4.

5. After that is the **Pre-load** sub-window (Figure 6.44d), which also has two options:
   - **Axial** (🔩) – recommended if the axial load on the bolt is known.
   - **Torque** (🔩) – used if the torque used to tighten the bolt is known. If the **Torque** option is checked, the program uses the **Friction Factor** (**K**, ⤙⤚) to calculate the axial force from the given torque. The following formulas are used:

$$F = \frac{T}{K*D} \text{ for a bolt with a nut and a torque applied on the nut}$$

$$F = \frac{T}{1.2*K*D} \text{ for a bolt without a nut and a torque applied on the head}$$

   where F is the axial force in the bolt, T is the applied torque, K is the friction factor and D is the major diameter of the shank.
   In our case, the axial load in the bolt is introduced (500 N).

Table 6.4
Formulas for Bolt Safety Checks

Loading	Formula	Notation
Axial load ratio	$R_A = \dfrac{F}{A_T * S}$	F – axial load calculated by the software A_T – tensile area S – strength value of connector's material
Bending load ratio	$R_B = \dfrac{D_n * M}{2 * S * I}$	M – bending moment calculated by the software D_n – nominal shank diameter I – area moment of inertia; it is calculated as $\quad I = \pi * r^4 / 4$
Shear load ratio	$R_S = \dfrac{V}{A_T * S}$	V – shear load calculated by the software
Combined load ratio	$(R_A + R_B)^2 + R_S^3$	
Pass/No Pass Safety Check		
Connector pass criterion	$\dfrac{1}{(R_A + R_B)^2 + R_S^3} > SF$	SF – user-defined factor of safety
Connector no pass criterion	$\dfrac{1}{(R_A + R_B)^2 + R_S^3} < SF$	

6. The last is the **Advanced Option** sub-window (Figure 6.44e).

- **Bolt series** are used when more than two components are bolted together. The cylindrical faces of solid bodies that are connected together (⊕) must be picked to mark the series.
- **Symmetrical bolt** is used if one or two planes of symmetry cut through the bolt. For one-half symmetry bolts, the plane or planar face of symmetry is selected in the **Reference Geometry** to be (◈ pink window). If symmetrical bolts are used, one-half or one-fourth of the total pre-load value and one-half or one-fourth of the total mass of the bolt according to the selected symmetry type are entered. Keep in mind that when listing the bolt forces after running the study, the results are equal to one-half or one-fourth of the total force.

The second fixture to be introduced to the model is the **Symmetry Fixture** (Figure 6.45):

Fixtures (⊞) → Advanced Fixtures → Symmetry

It defines the boundary conditions at the symmetrical plane and sets the translation normal to the selected face restrained.

Thus, all fixtures applied to the half of the unit are introduced.

The next stage is the introduction of the external loads. Three different types of loads will be input.

(a)   (b)

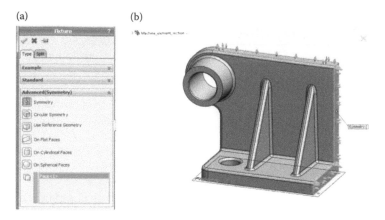

Figure 6.45
Defining the Symmetry Fixture. (a) Symmetry Fixture property manager. (b) Graphic area view when Symmetry Fixture property manager is active.

1. **Gravity** (⬤, marked with a big red arrow) is perpendicular to the **Top Plane**, and its value is equal to the earth's acceleration – 9.81 m/s² (Figure 6.46):

    External Loads → Gravity (⬤) → OK

2. **Pressure** (⊞, marked by small red arrows) models the joint between the studied machine unit and the shaft, which is pressed in the hole (Figures 6.42 and 6.47):

    External Loads → Pressure (⊞) → OK

3. **Torque** (⊞, marked by small magenta arrows) models the torsion of the shaft. It is important to keep in mind that the torsion of both shafts is counter-wise, anti-clockwise for the studied segment (Figures 6.42 and 6.48):

    External Loads → Torque (⊞) → OK

(a)   (b)

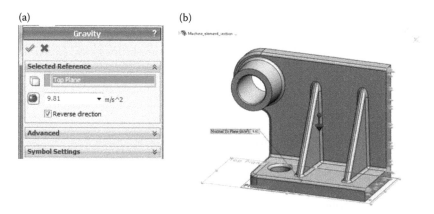

Figure 6.46
Defining the Gravity load. (a) Gravity property manager. (b) Graphic area view when Gravity property manager is active.

(a)   (b)

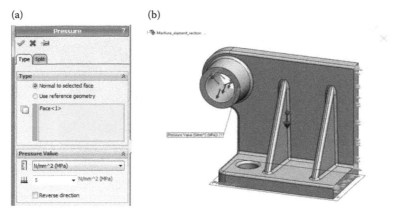

Figure 6.47
Defining the Pressure load. (a) Pressure property manager. (b) Graphic area view when Pressure property manager is active.

(a)   (b)

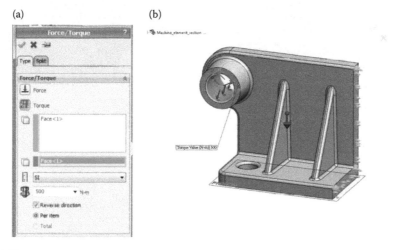

Figure 6.48
Defining the Torque load. (a) Torque property manager. (b) Graphic area view when Torque property manager is active.

The last stage in the development of the FE model is the creation of the mesh. A **Curvature-based mesh** with a maximal size of FEs of 5 mm and a minimal size of FEs of 1 mm is selected (Figure 6.49a and b). Each FE has 29 Jacobian points. The total number of the FEs is 73,617, whereas the total number of the nodes is 113,797. There are a few badly configured FEs, with an Aspect ratio of > 10 (Figure 6.49c). Therefore, mesh control (▣) at the hole faces is applied (Figure 6.49d and e). The new mesh consists of 109,384 FEs and 166,229 nodes. There are no FEs with Aspect ratio > 10 (Figure 6.49f and g).

The next stage is to run the model. The solver solves the system of about 355,350 equations, which equals the number of DOFs.

Some results of this study are given in Figure 6.50. In the context of the studied symmetrical boundary conditions, it is reasonable to pay special attention to the **UX** plot,

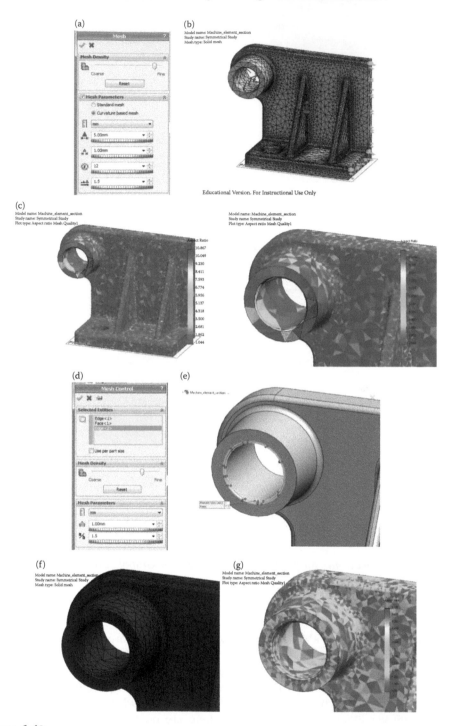

Figure 6.49

Meshing the model. (a) Mesh property manager. (b) Plot of the created mesh. (c) Aspect Ratio plot of the designed mesh. (d) Mesh Control property manager. (e) Faces to which mesh control is applied. (f) Mesh after Mesh Control is activated. (g) Aspect Ratio plot after Mesh Control is activated.

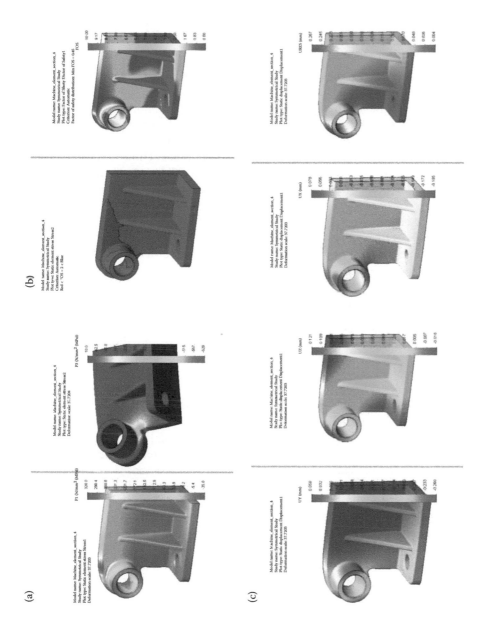

Figure 6.50
Results for symmetrical model with four ribs. (a) von Mises stresses plot. (b) Factor of safety plots. (c) Displacement plots.

admitting that UX = 0 for all nodes in the symmetrical face (Figure 6.50c). The most important conclusions about the obtained results are as follows:

- The extreme values of the tensile and the compressive stresses (Figure 6.50a) are far beyond the material strengths, and this is dangerous (maxP1 = 307 MPa compared to the tensile strength of the material of 151.7 MPa and minP3 = 628 MPa compared to 572 MPa compressive strength of the material).
- The minimal Factor of Safety is 0.45, and a large area of the material could not reach the indicative value of 2 FoS (Figure 6.50b).

As usual, the loads are pre-defined and could not be modified; a new design of this machine unit should be investigated.

The first idea that comes into a user's mind is to increase the number of the ribs and thus to make the construction less flexible. In fact, while studying this element, we see that the FoS problem is disposed at the area of the element, where the side cylinder component, which supports the shaft, joins the vertical plate. Thus, it is questionable whether the variation of the ribs will solve the problem. The CAD model was developed using the **LPattern** command; hence, it is easy to modify the machine unit by changing the number of ribs to six or two. As it is seen in Figure 6.51, the modifications of the ribs do not help to reduce the stress. Even more, the minimal FoS is reduced to 0.38. Thus, another solution must be searched.

Three different modifications of the unit's construction are discussed here:

- Increase in the outer diameter of the component that supports the shaft from φ28 mm to φ32 mm. The results are given in Figure 6.52a. The maximal displacement is decreased to 0.257 mm. The maximal tensile stress (maxP1) is decreased to 297 MPa, and the maximal compressive stress (minP3) is 577 MPa. These values are still larger than the tensile strength and the compressive strength of the material. Hence, a design modification of entirely different type must be discussed.
- To increase the thickness of the vertical plate from 8 to 16 mm. The results are given in Figure 6.52b. The maximal tensile stress (maxP1) is decreased to 190 MPa, and the maximal compressive stress (minP3) to 322 MPa. While the maximal tensile stress is still larger than the tensile strength of the material, the compressive strength is far below the limit. Thus, the area with the maximal tensile stress remains vulnerable. The maximal displacement is decreased to 0.14 mm, minimal FoS is 0.63 and the areas where the FoS coefficient is smaller than 2 are limited compared to all previous modifications.
- The last design modification presented here involves an increase in the outer diameter of the shaft support to 32 mm and a new design of the vertical plate, which thickness is set to 10 mm (Figure 6.52d). The extreme stresses are far below the strengths of the material, yet there are areas where the FoS is smaller than 2. This can easily be overcome by changing the radius of the fillet.

The conclusion is that the last design satisfies the initial safety requirements for the unit, made of Gray Cast Iron.

If the material of the machine unit is changed to **Alloy Steel**, the factor of safety significantly increases and its minimal value is 0.96. The areas, where FoS is below 2, are also limited (Figure 6.53).

Therefore, regarding the failure criterion, Alloy Steel for that machine unit can be a significantly better solution than Gray Cast Iron.

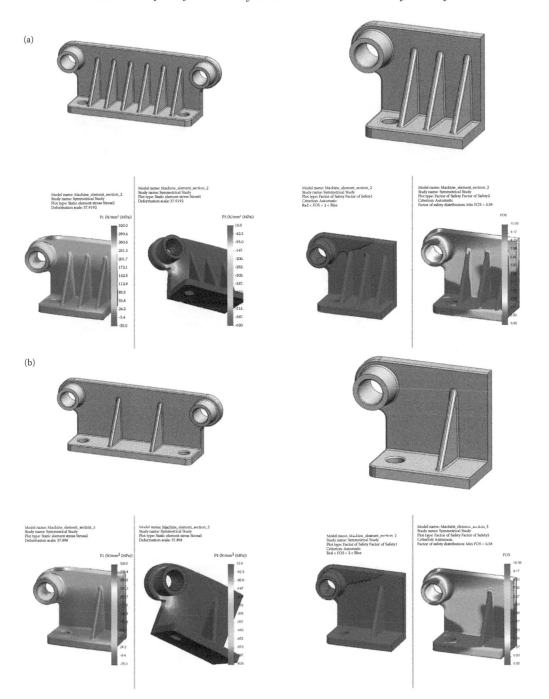

Figure 6.51

Results for symmetrical model with six or two ribs. (a) Modification of a machine element with 6 ribs: maxP1 = 313 MPa; minP3 = 600 MPa; max displacement = 0.266 mm; and minFoS = 0.39. (b) Modification of a machine element with 2 ribs: maxP1 = 318 MPa; minP3 = 615 MPa; max displacement = 0.266 mm; and minFoS = 0.38.

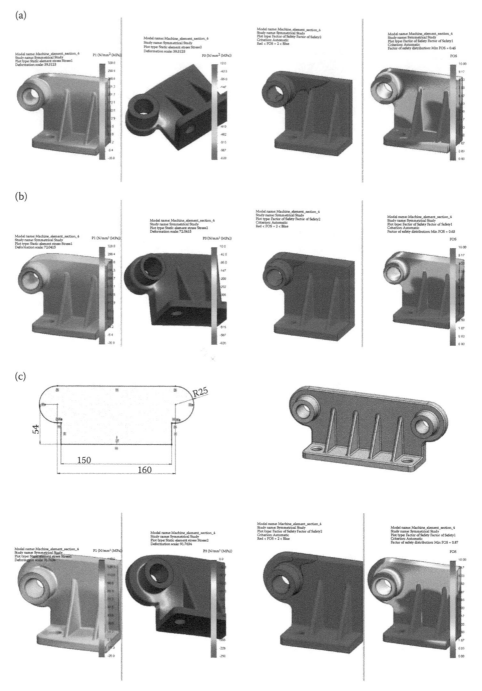

Figure 6.52
Results for machine elements with different variations, regarding its geometry. (a) Results for a machine element with a larger outer diameter of the shaft support: maxP1 = 295 MPa; minP3 = 577 MPa; max displacement = 0.257 mm; and minFoS = 0.45. (b) Results for a machine element with a thicker vertical plate: maxP1 = 190 MPa; minP3 = 322 MPa; max displacement = 0.144 mm; and minFoS = 0.63. (c) Results for a machine element with a new design of the vertical plate: maxP1 = 140 MPa; minP3 = 243 MPa; max displacement = 0.121 mm; and minFoS = 0.87.

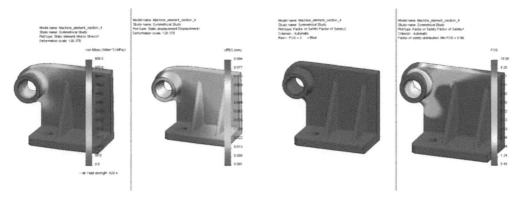

Figure 6.53
Results for a machine element made of Alloy Steel. maxP1 = 582 MPa (Yield Strength = 620 MPa), max displacement = 0.084 mm and minFoS = 0.96.

In this section, we studied a machine unit with planar symmetrical geometry, exposed to symmetrical boundary conditions.

We learned how to study a segment of a machine unit with planar symmetrical geometry, exposed to symmetrical loads and under symmetrical restraints, using the symmetry to simplify the model.

We learned how to apply

- Gravity load
- Symmetric boundary conditions using Symmetric Fixture
- Grounded Bolt fixture

We learned how to assess the stress plots when the object is made of brittle material and how to perform factor of safety calculations relying on the Mohr–Coulomb failure criterion.

We discussed a few ways to develop a structural optimisation targeting a reduction of stresses and an increase in the minimal FoS. These are

- Variations in design by adding more supporting elements
- Variations in design changing the sketch of some components
- Changing the material

CHAPTER 7

STATIC ANALYSIS OF A SHELL BODY

7.1   WHEN CAN AN OBJECT BE TREATED AS A SHELL? THIN OR THICK SHELL FEs? DIFFERENT APPROACHES FOR FEA OF A SHELL IN SW SIMULATION

In general, a body is considered a shell/plate when the contribution of shear deformations becomes non-significant and can be neglected. This happens when the ratio between the span of plate-bending curvature/the projected span of curvature and the shell thickness is larger than 10:1. According to some authors, this formulation itself is adequate and can be applied for a ratio down to 5:1 (https://wiki.csiamerica.com).

When studying shells, there are two basic formulations, depending on the inclusion of transverse shear deformation in plate-bending behaviour. These are the thin and the thick shell formulations. Thin-plate formulation neglects the transverse shear deformation, whereas thick-plate formulation does account for it. Thick-plate formulation has no effect on membrane (in-plane) behaviour, but only on plate-bending (out-of-plane) behaviour.

Shearing may become significant in locations of bending-stress concentrations, which occur near sudden changes in thickness or support conditions and near openings or re-entrant corners. Thick-plate formulation is best for such applications.

Thick-plate formulation is also recommended in general because it tends to be more accurate, though slightly stiffer, even for thin-plate bending problems in which shear deformation is truly negligible. However, the accuracy of thick-plate formulation is sensitive to mesh distortion and large aspect ratios, and therefore should not be used in such cases when shear deformation is known to be small.

The statement that thick shells tend to be stiffer than thin shells applies only to the bending components of shells and to models in which meshing is too coarse. When meshing adequately captures bending deformation, thick-shell elements are more flexible because of the additional shear deformation that is not captured through thin-shell formulation. Given pure-bending deformation, however, the thin-shell element is slightly more accurate; therefore, the thick-shell element may be stiffer for coarser meshes. This effect diminishes as the mesh is refined.

Stresses may be of greater concern than deflections. When shear deformation is expected to be important, the thick-shell FEs are recommended because they capture better the stress distribution. This is the case not only for thicker shells but also for regions near openings and other geometric discontinuities in which transverse shear deformation develops.

There are two ways to develop and study a shell body in SW Simulation and both will be discussed below.

- The first approach supposes the CAD model of the body to be developed using the **Surface** tool. During finite element analysis (FEA), the software automatically creates a shell uniting all surfaces. The type of the shell (thin or thick), its thickness and the off-set towards the surfaces are input by the user. This approach allows the use of composites and can generate shells with varying thickness. The input of contact constraint is necessary in this model.
- The second approach uses the **Sheet Metal** tool to develop the CAD model. While meshing the model, the software creates shell elements assuming that the thickness of the shell is equal to the thickness of the sheet metal. This approach is applicable for spatial forms with constant thickness.

This section discussed about when a body can be modelled and studied as a shell instead of studying it as a solid body. It briefly summarised and compared the two formulations of studying shells.

We obtained criteria to help us in deciding how to model and study the body – as a shell or as a solid body.

We learned about the advantages and disadvantages of the thin and the thick plate formulations. Therefore, we know how to justify our decision of which of the two formulations to use in the analysis.

7.2   DEVELOPMENT OF A CAD MODEL OF A SHELL USING SURFACE TOOL (Surface.sldprt)

The CAD model of the analysed object will be developed using the **Surface** tool. After starting a new part file (File → New → Part → OK) and setting the used units to '**millimetre-gram-second**' (Tools → Options → Document Properties → Units → Unit system MMGS → OK), the model is saved as **Surface.sldprt**. The next stage is to start the **Surface** tool and afterwards to create the model. The following stages must be done:

1. Activating the **Surface** tool

    After right clicking on the command bar, a pop-up menu appears (Figure 7.1b) and you must left click on **Surface** line (⬕, Figure 7.1b). The tools at the **Surface** toolbar are shown in Figure 7.1a and are defined in Figure 7.1c.

(a)

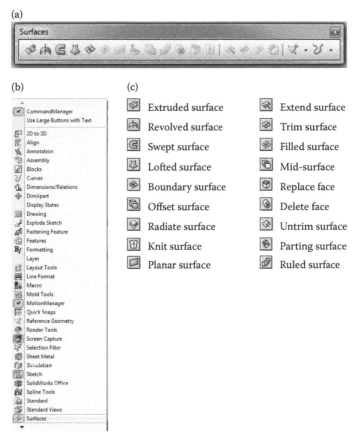

(b)

(c)

Extruded surface	Extend surface
Revolved surface	Trim surface
Swept surface	Filled surface
Lofted surface	Mid-surface
Boundary surface	Replace face
Offset surface	Delete face
Radiate surface	Untrim surface
Knit surface	Parting surface
Planar surface	Ruled surface

Figure 7.1
Activating Surface tool. (a) Surface toolbar. (b) Pop-up menu. (c) Tools for creating and providing surfaces.

2. Drawing **Sketch1** in **Top Plane** (Figure 7.2a).
3. Transforming **Sketch1** in a planar surface:

Surface → Planar Surface (⬚) → OK

Select **Sketch1** by clicking on it at the **Graphics area** (Figure 7.2b). As a result, its signature appears in the blue window of the **Planar surface** property manager (Figure 7.2c). Then click OK.
4. Defining a plane perpendicular to the existing planar surface and coinciding with its side edge (Figure 7.2d, e and f).

Features → Reference Geometry (⬚) → Plane (⬚) → OK

5. Drawing **Sketch2** in **Plane1** (Figure 7.3a).

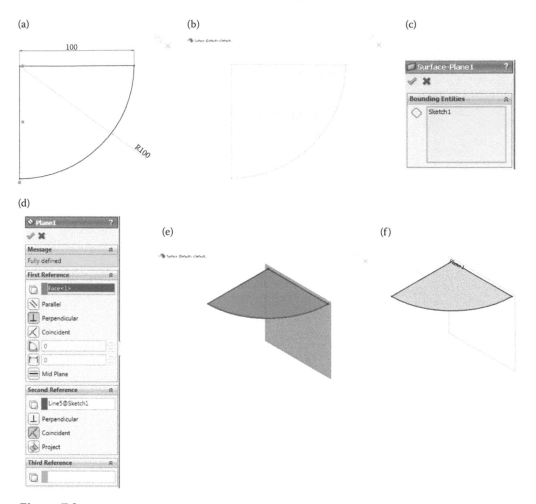

Figure 7.2

Creating a CAD model using Surface tool – stage from 2 to 4. (a) Sketch1. (b) Graphic area view while Planar surface property managed is active. (c) Planar surface property manager. (d) Plane property manager. (e) Graphic area view while Plane property managed is active. (f) New Plane1.

6. Transforming **Sketch2** into a planar surface (Figure 7.3b and c)

Surface → Planar Surface (▣) → OK

7. Defining **Plane2** at the side edges of both planar surfaces (Figure 7.2d and e).
8. Drawing **Sketch3** in **Plane2** (Figure 7.3f).
9. Transforming **Sketch3** into a planar surface (Figure 7.3g)

Surface → Planar Surface (▣) → OK

The ready CAD model is shown in Figure 7.3h.

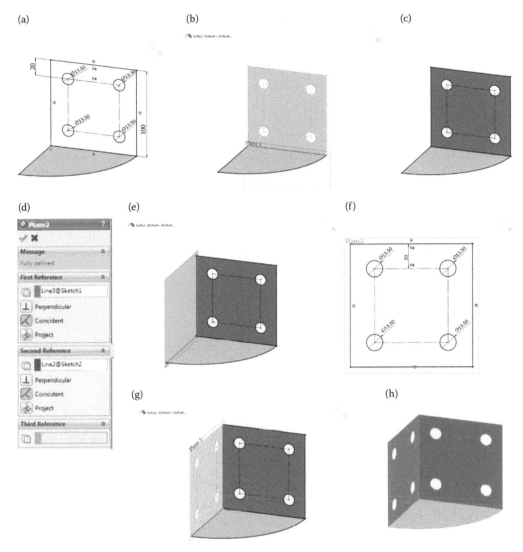

Figure 7.3
Creating the CAD model using Surface tool – stage from 5 to 9. (a) Sketch2 in Plane1. (b) Graphic area view while Planar surface property managed is active. (c) CAD model of two planar surfaces. (d) Plane2 property manager. (e) Graphic area view while Plane2 property managed is active. (f) Sketch3 in Plane2. (g) Transforming Sketch3 into planar surface. (h) The ready CAD model.

In this section, we modelled the studied object as a shell using the **Surface** tool. Thus, the software will automatically recognise this body as a shell and will perform the FEA using 2D FEs.

We learned

- How to start the **Surface** tool
- How to develop geometric models through it

7.3 FEA OF A SHELL, CREATED USING SURFACE TOOL (Surface.sldprt)

7.3.1 Pre-Processor Modelling of the Object

A new static analysis is started:

Simulations → New Study (⚘) → Static (🗗) → Study 1 → OK

The model unites three surface bodies (Figure 7.4a), in which material and thickness will be defined. The selected material is Alloy Steel and the thickness of the shell is set to 4 mm.

Right click on the name of the model in the **Simulation analysis tree** to open the pop-up menu. Start the **Apply Material** command (⸬).

Surface → Apply Material to All Bodies… (⸬, Material library open) →
SolidWorks Material → Steel → Alloy Steel → Apply → Close

Figure 7.4
Introducing the shell properties. (a) The surface bodies included in static Study 1. (b) Selecting all surface bodies after applying the material. (c) Shell Definition property manager. (d) Defined shell properties.

The quickest way to define the thickness of the shell bodies is to select all of them by clicking on their signatures in the **Simulation** analysis tree while the **Ctrl** button of the keyboard is pressed (Figure 7.4b). Then right click on them and pick **Edit Definition** from the pop-up menu. The **Shell Definition** property manager opens (Figure 7.4c).

The **Shell Definition** property manager is used to define the adopted calculation theory (thin-plate theory, where transverse shear deformations are neglected, or thick-plate theory). This is done by checking the correct button. As a general guideline, thin shells can be used when the thickness-to-span ratio is less than 0.05.

The shell can also be defined as a composite laminate. This case will not be discussed in the book.

The **Shell Definition** property manager defines the thickness of the shell elements. By default, the software assigns zero thickness to the shell geometry, but it can be modified by introducing a new value in the shell thickness window (🐚). In our case, this is 4 mm, which means that the thickness of the shell is 4 mm.

Based on the assumed shell thickness and the model geometry, the thickness-to-span ratio is less than 0.05. Thus, the thin plane theory can be adopted (Figure 7.4c).

Furthermore, there is the **Offset** option, which allows the user to control the position of the shell mesh relative to its surface. The option aligns the mesh to the top, the middle or the bottom faces of the shell. To align the mesh to a reference surface, an offset value is typed. By default, the mesh is always aligned to the middle face of the shell (🔲). The **Top Surface** (🔲) aligns the mesh with the top surface of the shell. The **Bottom Surface** (🔲) aligns the mesh with the bottom surface of the shell. For example, to model adjacent shells with differing thicknesses such that their bottom faces coincide, you can create the two surfaces and align them using the **Bottom surface** option. The **Specify Ratio** (🔲) aligns the mesh to a reference surface defined by an offset value that is a fraction of the total thickness. The offset value ranges between −0.5 and 0.5. The **Offset** option is not available for large displacement analysis of shells.

After the input of the material and the shell properties, the note 'Thickness not defined' disappears, and the chosen material is written for each surface body (Figure 7.4d).

The next stage is to define **Connections** (🔩). The program considers the three designed shells as free surface bodies despite the added geometric relations. Therefore, the type of the connections between them must be input manually (Figure 7.5a). **Edge Weld** connections are defined.

When defining a weld connection, the user chooses among the following types: **Fillet, Double-Sided** (🔺); **Fillet, Single-Sided** (🔺); **Groove, Double-Sided** (🔵); and **Groove, Single-Sided** (🔵). Depending on the selected weld type, the **Weld Type** sub-windows differ a little from each other (Figure 7.5b). The common input options include selection of the connected faces. **Face Set 1** (🔲) selects either the face of the shell or the face of the sheet metal body, which belongs to the terminated part. **Face Set 2** (🔲) selects the other face, which can belong to a shell, to a sheet metal object or to a solid body. For fillet welds, the selected faces of Sets 1 and 2 are perpendicular to each other. For groove welds, the selected faces are parallel. After picking the faces on the graphic area, their signatures appear in the windows on the left and the faces are coloured in blue (Figure 7.5c). The signatures of the **Intersecting Edges** (🔲) appear automatically. These edges are the touching edges between the selected **faces of Sets 1** and **2**, where the welding will be applied. Non-touching edges, which belong to the terminated part, can also be selected. The last option to be defined in that sub-window is **Weld Orientation** (🔳, 🔳). It is available only for single-sided welds. It provides the location of the weld with respect to the shell surface

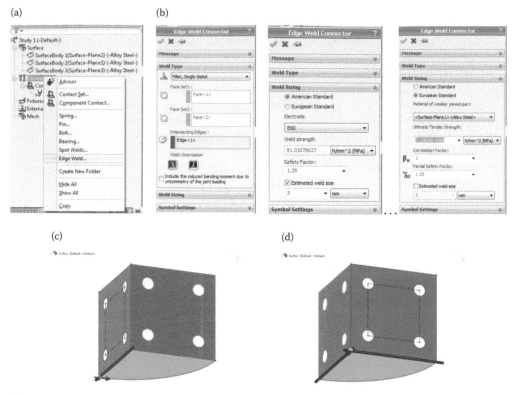

Figure 7.5
Defining the edge welds. (a) Connections pop-down menu. (b) Edge Weld Connector property manager for the first welding. (c) Graphic area view while defining the first welding. (d) Graphic area view while defining the second welding.

alignment. A red arrow in the graphics area shows the weld orientation (Figure 7.5c). When designing the weld parameters, the user must remember that due to asymmetry of the joint loading, the force, which is transmitted perpendicular to the longitudinal axis of the weld, can induce an additional bending moment at the weld throat. Considering or neglecting that moment is controlled through the **Include the induced bending moment due to asymmetry of the joint loading option** (not selected in the example, Figure 7.5b).

The next stage is determining the **Weld Sizing** (Figure 7.5b). The user can choose between **American Standard** (*American Welding Standard D1.1 and D1.2*) or **European Standard** (*Eurocode 3: Design of steel structures, Part 1.8: Design of joints, Section 4.5*).

The **American Standard** options include

- **Electrode** – to set the electrode's material.
- **Weld strength** – displays the selected electrode's material ultimate shear strength.
- **Safety factor** – reduces the allowable shear strength for the calculation of the weld strength. The allowable shear strength for the electrode's material is calculated as (Ultimate shear strength/Safety factor).
- **Estimated weld size** – let the program calculate the appropriate size for the weld connector.

The **European Standard** options include

- **Material of weaker joined part** – defines the weaker part, which is connected by the edge weld. The weaker part has the lesser material tensile strength.
- **Ultimate Tensile Strength** – displays the selected material's tensile strength.
- **Correlation factor** $\left(\hat{1}_{w}^{2}\right)$ – its value is between 0.8 and 1.0 for the weld calculations. The values for the correlation factor embedded in the software are from Eurocode EN 1993-1-8: 2002, Table 4.1.
- **Partial Safety Factor** $\left(\hat{1}_{1\alpha}^{3}\right)$ – the safety factor for joints is between 1.0 and 1.25. These values are from Eurocode EN 1993-1-8: 2002, Table 2.1.
- **Estimated weld size** – let the program calculate the appropriate size for the weld connector. It is determined according to Eurocode EN 1993-1-8: 2002, Paragraph 4.5.3.

European Standard is chosen for all defined edge welds in the case study, and the input options are given in Figure 7.5b. The **Graphics area** view, while the second edge weld is defined, is given in Figure 7.5d.

The next stage is the definition of the fixtures and loads. It is important to remember that the fixtures can be applied to the edges and vertexes of the shell. Since shell FEs have rotational degrees of freedom, there is difference between the **Immovable** (no translation) and the **Fixed Geometry** (no translation and no rotation) fixtures. **Immovable** fixtures at all edges around the holes are preferred over **Fixed Geometry** fixtures to replace the **Ground Bold** fixtures, assumed in the second example. The main reason for this choice is the relatively small pre-load tensile force in the bolds (Figure 7.6a)

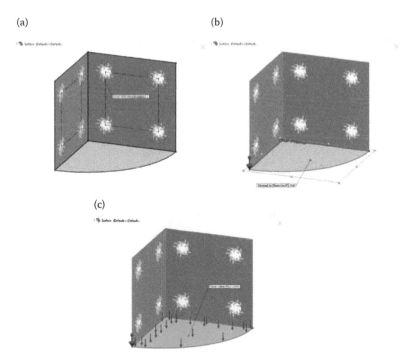

Figure 7.6
Defining the fixtures and the external loads. (a) Immovable fixtures at the edges of the holes. (b) Applied Gravity load. (c) Applied Force load.

The external loads at the analysed object are **Gravity** (⊙) and **Force** (⊥). **Gravity** is calculated based on the density value of the material and the thickness of the shell (Figure 7.6b). **Force** is equal to 10 kN and is normal to the horizontal surface (Figure 7.6c). Regarding the load, you must remember that for shells, the concentrated moments and forces can be applied using the **Force/Torque** property manager at faces, edges or vertexes. The pressure can be applied only to faces or to shell edges. The software uses the thickness of the shell and the length of the edge to calculate the equivalent force applied to the edge.

7.3.2 Meshing the Shell

After defining the connections, the fixtures and the loads, the model should be meshed. The program automatically determines the mesh to be of shell FEs.

Of course, any geometric model depending on the used commands and the way of creating the CAD design can be defined and analysed as a solid body model as well. Then, the software automatically uses tetrahedral solid elements. However, meshing of thin models with solid elements results in generating a large number of elements, since FEs of small element size are used. Just the opposite, using a larger element size deteriorates the quality of the mesh and leads to inaccurate results.

Surface models can only be meshed with shell elements. The shell mesh is generated on the surface, located at the mid-surface of the shell, unless nothing else is defined through the offset option.

By definition, shell elements are 2D elements capable of resisting membrane and bending loads. They can be either linear triangular elements (or first-order elements; Figure 7.7a) or parabolic triangular elements (or second-order elements; Figure 7.7b). Three corner nodes, connected by three straight edges, define a linear triangular shell element. Three corner nodes, three mid-side nodes and three parabolic edges define a

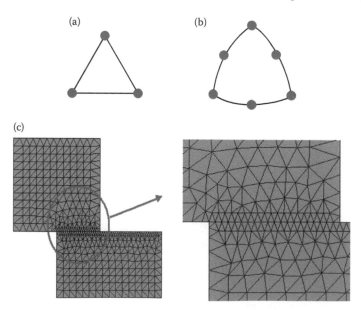

(a)   (b)

(c)

Figure 7.7
Shell finite elements (SW Simulation Help). (a) Linear triangular element. (b) Parabolic triangular element. (c) Compatible mesh along shells sharing edges.

parabolic triangular element. For structural studies, each node in shell elements has six degrees of freedom: three translations and three rotations. The translational degrees of freedom are motions in the global X, Y and Z directions. The rotational degrees of freedom are rotations about the global X, Y and Z axes. When using shell elements, the software generates one of the following types of elements depending on the active meshing options for the study: **draft-quality mesh**, when the automatic mesher generates linear triangular shell elements; or **high-quality mesh**, when the automatic mesher generates parabolic triangular shell elements.

The software generates a continuous mesh on shells sharing edges. Meshing generates a compatible mesh along the interface and merges the nodes automatically regardless of mesh control and contact settings (Figure 7.7c). If a mesh control is applied to one of the shared edges, the software uses the smaller size of FEs for both edges. Mesh controls can be applied on the appropriate faces, edges and vertices. Remember that, by default, the software assigns a thin shell formulation to each surface body. The user can edit it before running the study. A reasonably fine draft-quality mesh gives results that are generally similar to the results obtained from a high-quality mesh with the same number of elements. The difference between the two results increases if the model includes curved geometry.

A mesh of the following options is generated: curvature-based mesh, Max FE size – 4 mm; Min FE size – 1 mm. Draft Quality option is checked (Figure 7.8a and b). The

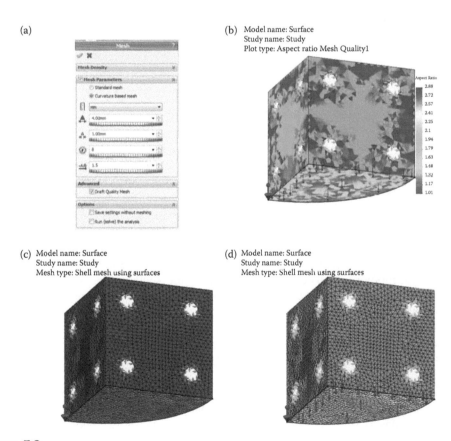

Figure 7.8
Meshing of the shell. (a) Mesh property manager. (b) Aspect Ratio plot. (c) Inner surface of the shell is 'bottom'. (d) Inner surface of the shell is 'top'.

bottom faces' colour is defined to be orange. If you like, you can change it through the following path of commands (Figure 4.1b):

Simulation → Options → System Options → General → Mesh colors → Shell bottom face color → Orange → OK

Each surface can be flipped using the **Flip Shell Elements** command, which is accessible through

Mesh (right click) → Flip Shell Elements

Thus, you can define the inner surface of the shell to be 'bottom' (Figure 7.8c) or 'top' (Figure 7.8d).

7.3.3 Viewing the Results

There are no significant differences in viewing the results of the FEA of shells compared to those of solid bodies. The software generates data about displacements, strain and stresses. The new points that will be discussed in this chapter concern the different types of stresses, in relation to the deformations they are related to, and the displacement to the mid-plane of the shell. The following stresses can be viewed:

- **Top** – provides the total stresses (bending + membrane) at the top face of the shell
- **Bottom** – provides the total stresses (bending + membrane) on the bottom face of the shell
- **Membrane** – provides membrane stress component
- **Bending** – provides bending stress component

Before continuing with plotting the results, some additional explanation on the topic is provided. It is of great importance to choose the correct 'top/bottom' side while doing shell calculations. Usually the top side lies above the mid-plane of a horizontal plate, and its offset distance is one-half of the thickness of the plate (Figure 7.9a). The stresses that appear in the shell have two main components: membrane stresses and bending stresses. Membrane stresses are related to the in-plane deformations of the shell. They are due to the tensile/compressive forces and are constant along the thickness of the shell. Bending stresses are related to the out-of-plane deformations of the shell. Bending moments or forces perpendicular to the shell mid-plane cause them. It is assumed that these stresses distribute linearly along the thickness of the shell; their extreme values, that is, maximal tensile and maximal compressive stresses, are at the two layers most distant from the mid-plane. They are equal, while the mid-plane bending stress is zero. The total (normal) stresses are a sum of the membrane and of the bending stresses. Thus, their values at the top layer differ from the values at the bottom layer (Figure 7.9b).

It is important to notice that this is the simplest explanation of the nature of these stresses. It is true for linear objects and it is fundamental for the mechanics of materials, which studies 1D objects (bars, beams or frames). Therefore, when trying to find any similarity between the given explanation and the one generated by the software results, you have to remember that software calculates the stresses in a way based on the above explanation, but this time including more factors and the fact that the shells

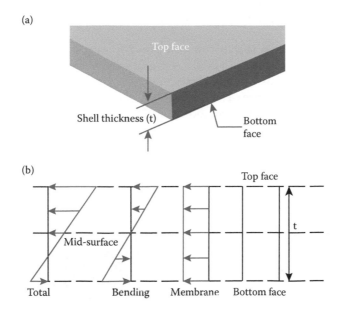

Figure 7.9
Stresses in a shell (SW Simulation Help). (a) Shell view. (b) Vertical distribution of the stresses.

are calculated as 2D objects. Additionally, not all maximal values rise in the same element (Figure 7.10c and d). Some of the obtained plots for a 4-mm shell and an assumed thin formulation are shown in Figure 7.10, where you can compare the membrane and the bending stress distributions across the shell.

The extreme values of some results are given in Table 7.1. The Yield Strength of the chosen material (Alloy Steel) is 620.4 MPa. It is seen that the minimal value of FoS is smaller than 1, which is unallowable in this case. More important is the fact that this low value is due to the calculation process. If you look at the given picture (Figure 7.10f, right), you will see that the vulnerable area involves a few FEs only. Even more, if you change the formulation and assume thick-plate formulation (which is considered to be more accurate but more complicated and which requires more computer resources and calculation time), you will see that the discrepancies between the stresses in the element and in the node mode almost disappear and the FoS rises to 1.06. Therefore, we can conclude that the FoS discrepancy is due to numerical calculations, including shell formulation and mesh definition. Even more, if the target value of FoS is 1, we can assume to keep the object design as it is.

We performed an FEA of a shell object, modelled through the **Surface** tool. The software automatically recognised the surfaces as shells and uses 2D FEs to mesh them.

The user inputs the thickness of each surface and assumes the formulation for calculations (a thin or a thick plate) and the displacement of the shell to the surface (the offset).

It is important to remember that despite the existing geometric mates between the surfaces, contacts must be applied. We chose edge welding and we define two welded edges.

The limitations concerning the application of the fixtures and of the loads are explained.

2D FEs and the meshing of shells are discussed. It is explained why it is better to use 2D elements for shells instead of 3D FEs, and why we do not model and prefer not to analyse the shells as solid objects.

The nature of the four stresses calculated by the software (top, bottom, membrane and bending) was explained and their plots were presented.

(a) Model name: Surface
Study name: Study 1
Plot type: Static element stress (Top) Stress1
Deformation scale: 1

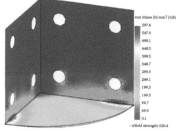

(b) Model name: Surface
Study name: Study 1
Plot type: Static element stress (Bottom) Stress1
Deformation scale: 1

(c) Model name: Surface
Study name: Study 1
Plot type: Static element stress (Membrane) Stress1
Deformation scale: 1

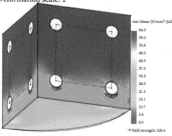

(d) Model name: Surface
Study name: Study 1
Plot type: Static element stress (Bending) Stress1
Deformation scale: 1

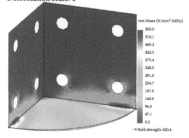

(e) Model name: Surface
Study name: Study 1
Plot type: Static displacement Displacement1
Deformation scale: 1

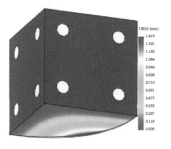

(f) Model name: Surface
Study name: Study 1
Plot type: Factor of Safety Factor of Safety 1
Criterion: Automatic
Factor of safety distribution: Min FOS = 0.97

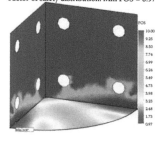

Figure 7.10

Plots for a 4-mm shell, a thin formulation assumed, shell thickness – constant. (a) Top von Mises stresses, presented in element mode, t = 4 mm. (b) Bottom von Mises stresses, presented in element mode, t = 4 mm. (c) Membrane von Mises stresses, presented in element mode, t = 4 mm. (d) Bending von Mises stresses, presented in element mode, t = 4 mm. (e) Displacement plot, t = 4 mm. (f) FoS plot, t = 4 mm.

Table 7.1
Results for Analysed Shell for Thin and Thick Shell Formulations

Surface Shell, with Thickness of 4 mm	Maximal von Mises Stresses (MPa)				Displacement (mm)	FoS
	Top	Bottom	Membrane	Bending		
Thin-Plate Formulation						
Element mode	597.6	529.0	64.3	563.0	1.419	0.97
Node mode	637.4	569.6	80.7	603.2		
Thick-Plate Formulation						
Element mode	585.08	520.0	62.8	550.5	1.462	1.06
Node mode	583.3	517.1	72.4	549.9		

We learned

- How to define shell properties through the **Shell Definition** property manager and how easy the shell thickness and the assumed calculating formulation can be changed
- How to define a welding, what the different types of welding are and the standard definitions used by the program
- How to apply fixtures and loads at shells and what the main differences are with solid body applications
- How to define a mesh of 2D FEs and how to flip their sides
- What is the nature of top, bottom, membrane and bending stresses and how to plot them

---

7.4 DEVELOPMENT OF A CAD MODEL OF A SHELL USING SHEET METAL TOOL (Sheet_Metal.sldprt)

In this section, the CAD model of the same part will be developed by the Sheet Metal tool and following further instructions.

The stages to be fulfilled are as follows:

1. Starting **a new model**:

$$\text{File} \rightarrow \text{New} \rightarrow \text{Part} \rightarrow \text{OK}$$

Save file as **Sheet_Metal.sldprt**.

2. Setting the options:

$$\text{Tools} \rightarrow \text{Options} \rightarrow \text{Document Properties} \rightarrow \text{Units} \rightarrow \text{Unit system MMGS} \rightarrow \text{OK}$$

3. Drawing **Sketch1** in the **Top Plane** (Figure 7.11a).
4. Extruding **Sketch1** (Figure 7.11b and c):

$$\text{Features} \rightarrow \text{Extrude Boss/Base} (\text{⬚}) \rightarrow \text{OK}$$

Select the sketch. Let the feature be extended as **Blind** to 100 mm.

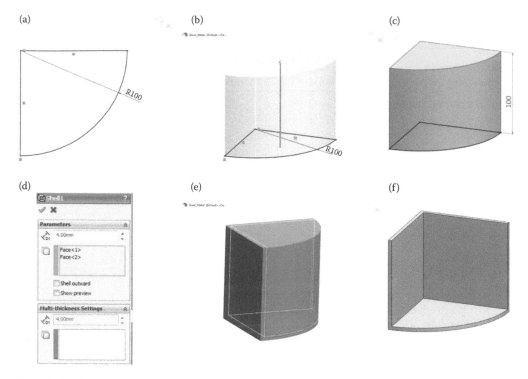

Figure 7.11
*Development of the CAD model of the Sheet_Metal part – stages 3 to 5. (a) Drawn Sketch1.
(b) Extrusion of the sketch. (c) The modelled solid body after the extrusion of Sketch1. (d) Shell
property manager. (e) Preview of the hollowed part. (f) Created shell.*

5. Defining the shell geometry from the modelled solid body:

Features → Shell (▣) → OK

The **Shell** tool hollows out a part, leaves open the selected faces and creates thin-walled features on the remaining ones. If no face on the model is
selected, the software creates a closed, hollow model. The program can also
create a shell model using multiple thicknesses. To create a shell feature of uniform thickness, we set the thickness of the kept faces (⌀) in the **Parameters**
sub-window and select one or more faces in the **Graphics area** for **Faces to
remove** (⬚). The **Shell outward option** increases the outside dimensions of
the part. The options of the **Shell** property manager include the introduced
thickness of the shell, which is set to 4 mm, and the pointed faces, which are
coloured in blue (the top face as **Face1** and the cylindrical face as **Face2**,
Figure 7.11d and e). The user can see the preview of the hollowed part before
confirming the execution of the command. The created shell is shown in
Figure 7.11f.
6. Drawing **Sketch2** in the outer plane of the vertical side of the shell (Figure 7.12a).
To draw the sketch, the **Offset** command is used to set the construction
lines and to locate the centres of the circles. The diameter of each circle is 13.5
mm (Figure 7.12b).

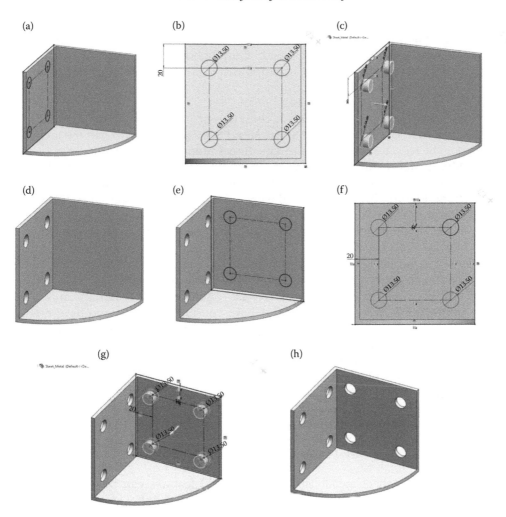

Figure 7.12
Development of the CAD model of the Sheet_Metal part – stages 6 to 9. (a) Displacement of Sketch2. (b) Sketch2. (c) Picked contours to be cut-extruded. (d) Cut-extruded holes. (e) Displacement of Sketch3. (f) Sketch3. (g) Selected to be cut contours. (h) The ready geometry of the shell.

7. Cutting the holes in the side:

Features → Extruded Cut (⬛) → OK

Select the regions, outlined by the circles, as shown in Figure 7.12c. Pay attention to their input in the **Selected Contour** sub-window by picking them in the **Graphic area**. The **distance** (⬛) of the extrusion is set to 10 mm as this is supposed to be the thickest studied sheet metal (Figure 7.12d).

8. Drawing **Sketch3** in the outer plane of the other vertical side of the shell (Figure 7.12e and f).

9. Cutting the holes in the second side (Figure 7.12g and f):

Features → Extruded Cut (⬛) → OK

If we stop modelling the part now and start performing FE analysis, despite the used **Shell** command and the created shell geometry, the software treats the part as a solid body and uses solid FE throughout the calculations.

Therefore, the part must be transformed to a shell. This will be done through the **Sheet Metal** tool.

10. Activating the **Sheet Metal** toolbar.

After right clicking on the command bar, a pop-up menu appears (Figure 7.13b). The user have to left click on the **Sheet Metal** (⊞, Figure 7.13a) to start the tool. Among the most useful commands in the **Sheet Metal** toolbar are as follows:

- **Sheet Metal** (⊞) – contains the default bend parameters. It edits the default bend radius, bend allowance, bend deduction or default relief type.
- **Base-Flange** (⬥) – represents the first solid feature of the sheet metal part.
- **Flat-Pattern** (⬒) – flattens the sheet metal part. By default, it is suppressed, as the part is in its bent state. Unsuppressing the feature will flatten the sheet metal part.

11. Converting the shell part to a sheet metal.

SolidWorks provides three ways to create a sheet metal part:

- Convert a solid part to a sheet metal part – solid or surface bodies can be converted to sheet metal parts that have constant thickness and have no shells or fillets, have either a shell or fillets or have both a shell and fillets
- Create the part as a sheet metal part using sheet metal-specific features
- Build a part, shell it and then convert it to sheet metal.

To model the studied unit, the third way is selected.

A shell part of uniform thickness is already created (Figure 7.12h), and the **Sheet Metal** toolbar is activated (Figure 7.13a).

We start a **Convert to Sheet Metal** command (⬛, Figure 7.13):

Sheet Metal toolbar → Convert To Sheet Metal

After creating the sheet metal part, all sheet metal features can be applied to it. Through the **Convert to Sheet Metal** property manager (Figure 7.13c), the user can specify the fixed face and thickness of the sheet metal part, the default bend radius and the edges or fillet faces on which to create bends. If an edge already has a fillet applied, the radius of the fillet is used as the bend radius for the new sheet metal part. The software automatically selects the edges on which rips are applied. However, this can also be done manually by selecting rip edges and using rip sketches.

The options of the eight sub-windows of the **Convert to Sheet Metal** property manager are given as follows (Figure 7.13d and e):

- **Sheet Metal Gauges** – This option is available only the first time the **Convert to Sheet Metal** tool is activated. It enables selecting a gauge table as the base of the sheet metal feature. The sheet metal parameters (material thickness, bend radius and bend calculation method) use the values stored in the gauge table unless something else is overridden. The table can be selected (⬛) through the browse option, which directly opens a file path to the software directory **Sheet Metal Gauge Tables** or through

Tools > Options > System Options > File Locations

and selecting **Sheet Metal Gauge Table** in the **Show folders for** window.

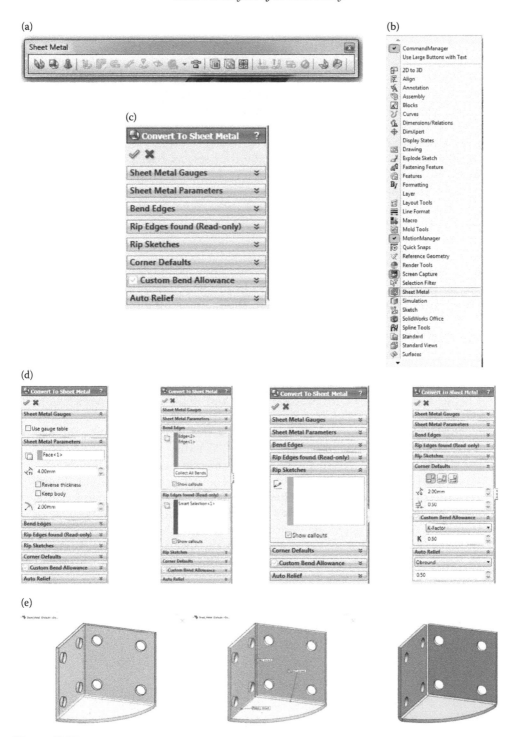

Figure 7.13
Converting the solid part to a sheet metal. (a) Sheet Metal toolbar. (b) Activating Sheet Metal menu. (c) Convert to Sheet Metal property manager. (d) Options of the tabs of Convert to Sheet Metal property manager. (e) Graphic area views when Convert to Sheet Metal property manager is open.

- **Sheet Metal Parameters** – At first, we select a fixed entity (⬜): this is the face that remains in place when the part is flattened – the bottom face in the horizontal plate (the green face in Figure 7.13e, the figure on the left). **Sheet thickness** (⬦) is 4 mm. **Reverse Thickness** changes the direction in which the sheet thickness is applied – *not checked*; **Keep body** keeps the solid body to use with multiple **Convert to Sheet Metal** features or designates that the entire body be consumed by the sheet metal feature – *not checked*; **Default radius for bends** (⌒) is set to 2 mm.

- **Bend Edges** – Through that option, the user selects edges/faces that represent bends (⬜). It is enough to click on an edge in the **Graphics area** to add it to the list of bend edges in the pink window – *click on the pink edges* (the pink lines in Figure 7.13e, the figure on the middle). **Collect All Bends** is pushed when there are pre-existing bends to find all of the appropriate bends in the part – *no pre-made bends*. When checked, **Show callouts** displays callouts in the **Graphics area** for all selected edges.

- **Rip Edges found (Read-only)** – When selecting a bend edge, the corresponding rip edge is automatically displayed in the violet window (Figure 7.13d). **Show callouts** displays callouts in the graphics area for all rip edges (Figure 7.13e, the figure on the middle).

- **Rip Sketches** – This option selects a sketch to add a rip (⬓); a 2D or a 3D sketch is selected to define the required rip (its signature immediately appears in the green window). **Default gap for all rips** (⬦) should be introduced. No rip sketches are defined for our model.

- **Corner Defaults** – The settings apply to all rips in the **Graphics area** whose callouts say 'Default'. These defaults can be overridden by setting new options for individual rips in the same callouts. At first, the user should define the **Rip Type** by clicking on one of the icons (⬓, ⬓, ⬓); then the user should define the **Rip Width** (⬦) and finally the **Default overlap ratio for all rips** (⬓). **Open butt** (⬓) is selected and the **Default gap for all rips** (⬦) is 2 mm. **Default overlap ratio for all rips** (⬓) adjusts the material lengths and is taken into account only for **Overlap** (⬓) and **Underlap** rips (⬓). Because of that, its value is of no importance to our model.

- **Custom Bend Allowance** – It is active only the first time when the **Convert to Sheet Metal** tool is used. **Bend Allowance Type** – K-factor; **Bend Allowance Value** – 0.50.

- **Auto Relief** – The software automatically adds relief cuts where needed when inserting bends. The type of the relief cut could be **Rectangular, Obround** or **Tear**. If you select **Rectangular** or **Obround** options, a relief ratio needs to be introduced. The relief ratio is calculated as

$$\text{Relief ratio} = \frac{d}{\text{part thickness}}$$

where the distance d represents the width of the auto relief cut and the depth by which it extends past the bend region. The value of the relief ratio must be between 0.05 and 2. The higher the value, the larger the size of the relief cut added during the insertion of bends. For our model, the inputs options are type: Obround and value: 0.50.

After closing the **Convert to Sheet Metal** tool, three new lines automatically appear in the **FeatureManager** design tree. They hold the properties of the designed sheet metal.

- **Sheet Metal1** (⊞, Figure 7.14a): If we open it and click on the **Edit Feature** icon (▤), we will see the properties of the designed sheet metal:
    - **Fixed Face or Edge** (▤), which can be modified through that window
    - **Bending radius** (⌒), which can also be modified through that property manager
    - **Thickness of the sheet** (↖ᴛ), which cannot be changed through the **Sheet Metal** property manager
- **Convert – Solid1** (▦) – This line unites all properties introduced to the model through the **Convert to Sheet Metal** property manager. All input values and properties, including the sheet metal thickness, can be modified through that feature. As sub-directories of **Convert – Solid1,** two bending lines appear: **Sharp Bend1** and **Sharp Bend2** (▥). They represent both bended edges (Figure 7.14b).
- **Flat-Pattern1** (▣), which is suppressed by default. If you unsuppress (↥) it, you will see the flattened sheet (Figure 7.14c).

When creating a multibody sheet metal part, in the **FeatureManager** design tree, the main tree lists each body and its features in the order in which they are added. In addition, the cut list contains a separate representation for each body, with a flat pattern specific to the body. By default, the **Automatic** option of the cut list is active; correspondingly, the **Update** option is added to the right-click menu. There are two icons that show the update information about the generated cut list: ▦ (indicates that the cut list needs to be updated) and ▦ (indicates that the cut list is up-to-date).

This time, we built the unit from the previous section as a solid body, shelled it and then converted it to sheet metal using the **Sheet Metal** tool.

(a)   (b)   (c)

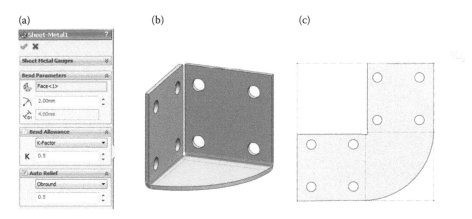

Figure 7.14
Final data for the sheet metal shell. (a) Sheet metal property manager. (b) Both bended edges. (c) Flattened sheet.

We learned

- How to use the **Shell** feature to hollow a solid body.
- How to convert a solid part to a sheet metal using the Sheet Metal tool.
- The properties of the **Sheet Metal** tool.
- How to modify the model properties.
- How to obtain the flattened sheet model, etc.

7.5   FEA OF THE SHELL, CREATED USING SHEET METAL TOOL (Sheet_Metal.sldprt)

After the start of the new static study (Simulation → New Study → Static → OK), the software automatically recognises the sheet metal structure as a shell structure made of 2D FEs. Even more, it automatically extracts and assigns the thickness of the sheet metal to the shell. The thickness cannot be modified through the **Shell Definition** command. Only the calculating mode – thin or thick plate formulation – can be changed. The thickness of the shell is modified by the **Convert to Sheet Metal** tool at the geometric model. Another option is the transformation of the shell to a solid body. This can be done by right clicking on the **Sheet_Metal** (◇) and selecting **Treat as Solid** from the pop-up menu. If so, the software will use 3D FEs when meshing the body. Right now, we want to avoid this procedure.

Compared to the shell model developed through the **Surface** tool, this time, the software automatically recognises existing connections, in compliance to the flattened sheet geometry (Figure 7.14c). There is no need to introduce additional connections as it was in the previous studied model.

To make it easier to compare the results of the two studied models, again, the applied material is Alloy Steel and equivalent loads. A vertical force of 10 kN is applied to the top horizontal surface (pink arrows). The gravity is also considered (the red arrow in Figure 7.15a).

Regarding the fixtures and the external loads acting at the body, you will notice that they can be applied to the cross faces of the sheet metal as well. The software automatically transfers them to the mid-surface. Thus, to apply a fixture or a load to a

(a)   (b)

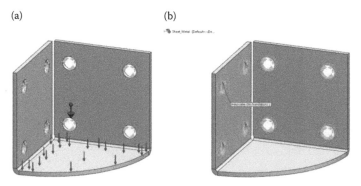

Figure 7.15
Applied restraints at the model. (a) Applied loads. (b) Applied Immovable fixtures.

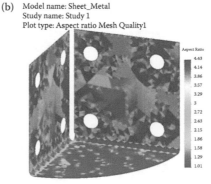

(a) Model name: Sheet_Metal
Study name: Study 1
Mesh type: Shell mesh using mid-surfaces

(b) Model name: Sheet_Metal
Study name: Study 1
Plot type: Aspect ratio Mesh Quality1

Figure 7.16
Meshing the model. (a) Plot of the mesh. (b) Plot of the Aspect Ratio.

shell edge, the associated face of the solid must be selected. To apply a fixture or load to the shell vertex, the associated edge of the solid must be selected. This enlarges the scope of applicable restraints. For example, **Grounded Bolt** fixtures can be applied at the hollows. We will do this and then will compare the results. **Grounded Bolds** with a diameter 13.5 mm and an axial force of 500 N suit quite well to that problem. The calculated results for the case of **Ground Bold** fixtures with the described properties are as follows: FoS = 1.09; max displacement 1.33 mm; maximal von Mises top stresses (element mode) = 560 MPa.

Further, we will continue with the case when **Immovable** fixtures at the cylindrical faces of the hollows are applied (Figure 7.15b).

When meshing the shell, the software automatically uses 2D FEs. It extracts the mid-surfaces and generates shell mesh at the mid-surface. The **Flip Shell Elements** command is available. The program recognises the bended edges (Figure 7.14b) as separate shell components (Figure 7.16).

There are no tricky points in the solution. Some final plots and values are provided in Table 7.2. They compare the 'sheet metal' shell with **Immovable** fixtures to that with **Grounded Bold** fixtures. The plots of the stresses, of the resultant displacement and of FoS for both cases with equal ranges of the used charts are displayed in Figure 7.17.

Based on the provided values and plots, we can conclude that the use of **Immovable** fixtures makes the model steadier and causes larger stresses and displacements compared to the use of **Grounded Bold** fixtures. The **Grounded Bold** fixture ensures more accurate stress distribution and a FoS that is a little bit higher.

Table 7.2

Comparison of the Results for the 'Sheet Metal' Part, Depending on the Applied Fixtures

Fixture	Max von Mises Stresses (Element Mode) (MPa)				Max Displacement (mm)	Min FoS
	Top	Bottom	Membrane	Bending		
Immovable	611.53	496.23	82.81	545.67	1.39	1.01
Grounded Bold	560.25	480.32	67.73	507.01	1.33	1.01

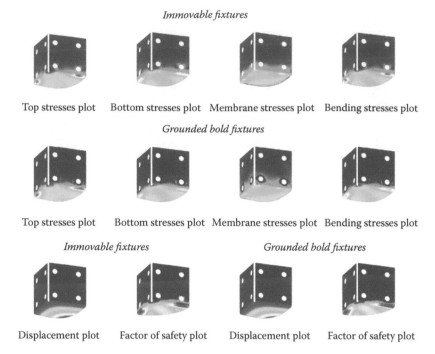

Figure 7.17
Results of both FEAs.

In this section, we made an FEA of a shell, whose geometry is developed using the **Sheet Metal** tool. We compared the impact of the applied Immovable fixture to the applied Ground Bold fixture on extreme stresses and displacements.

We learned how to perform FEA of a shell, whose geometry is designed using the **Sheet Metal** tool. We compared

- **Shell Definition** options for shells of uniform thickness, whose geometry is developed in both different ways, and learned how to modify its thickness, offset disposition and bending radii
- The processes of applying fixtures and loads and the limits imposed by different geometry modelling

7.6   COMPARISON OF THE RESULTS FROM THE TWO CASE STUDIES

The results of the two previously discussed cases will be compared, and some comments and recommendations on the selection of the approach of geometry modelling – **Surface** tool versus **Sheet_Metal** tool – will be given.

To do so, the results of both previous case studies (material: Alloy steel; thickness of the sheet: 4 mm; fixtures: Immovable; loads: vertical force of 10 kN and gravity) are systematised and analysed.

The first questions to be answered are as follows: What will happen to the stresses and the displacements if we model the studied object as a solid body? What is the impact of the use of a different type of FEs? Based on the given data (Table 7.3), we can conclude that the use of solid FEs provides minimal FoS and maximal stresses. Thus, it ensures the highest level of safety, but its use is not recommended because of the great demand for computer resources and time. At that point, we do not discuss whether thin- or thick-plate formulation is used for the calculations given above.

The next set of questions to be answered is as follows: What is the impact of the choice of formulation – thin or thick plate? Can a wrong choice cause significant errors? The data provided in Table 7.4 compare the results of the FEA of the **Surface. sldprt** adopting both formulations. As the thickness increases, the discrepancies, calculated towards the 'thick-plate' values, also increase. Thin-plate formulation provides larger stresses, resultant displacements and smaller FoS. Thus, it is safer to use it.

The last item to be discussed is the comparison of the results of FEA of the two models – **Surface.sldprt** versus **Sheet_Metal.sldprt**. The data given in Table 7.5 are obtained according to thick-plate formulation. The most important conclusion, which can be done regarding the data, is that it is not correct to compare the results and to make any recommendations. More important is the nature of these results. Vulnerable zones are located along the bended edges of the sheet metal model. Thus, the radii of the bends influence strongly the maximal stresses, and their least modification can significantly reduce or increase the stresses. The given data are for a bending radius equal to 5 mm. The **Surface.sldprt** has adopted edge welding, which ensures less deformable edge connection, and the stresses distribute more uniformly. The only recommendation I can make, based on all the given data, is as follows: If I had to choose which of the two supporting elements to produce, I would choose to make the support made of three welded surfaces because of the higher level of safety reached. The manufacturing operations and their cost were not considered when I made my decision.

Table 7.3
Comparison of Solid Body Results to Shell Model Results

			Shell Model, Developed by	
Cases to be Compared		Solid Body Model	Surface Tool	Sheet Metal Tool
Max von Mises stresses (MPa)	Top	665.14	597.6	611.53
	Bottom		529.0	496.23
	Membrane		64.3	82.81
	Bending		563.0	545.67
Displacement (mm)		1.37	1.42	1.39
FoS		0.69	0.97	1.01

Table 7.4

Comparison of Thin-Plate versus Thick-Plate Results

Used Formulation	Maximal von Mises Stresses (MPa)				Displacement (mm)	FoS
	Top	Bottom	Membrane	Bending		
Shell Thickness t = 2.5 mm; F = 3 kN						
Thin plate	454.0	418.7	36.0	436.3	1.692	1.28
Thick plate	447.6	414.3	35.4	429.4	1.717	1.39
Discrepancy	1.41	1.05	1.67	1.58	−1.48	−8.59
Shell Thickness t = 3 mm; F = 5 kN						
Thin plate	528.0	480.0	46.9	503.9	1.646	1.10
Thick plate	519.6	473.9	46.0	494.8	1.679	1.20
Discrepancy	1.59	1.27	1.92	1.81	−2.00	−9.09
Shell Thickness t = 3.5 mm; F = 5 kN						
Thin plate	390.1	350.4	37.4	370.1	1.059	1.49
Thick plate	383.4	345.2	36.5	362.8	1.086	1.63
Discrepancy	1.72	1.48	2.41	1.97	−2.55	−9.40
Shell Thickness t = 4 mm; F = 10 kN						
Thin plate	597.6	529.0	64.3	563.0	1.419	0.97
Thick plate	585.08	520.0	62.8	550.5	1.462	1.06
Discrepancy	2.10	1.70	2.33	2.22	−3.03	−9.28
Shell Thickness t = 5 mm; F = 10 kN						
Thin plate	383.8	332.1	50.1	357.6	0.750	1.52
Thick plate	375.0	325.2	49.1	348.5	0.781	1.66
Discrepancy	2.29	2.08	2.00	2.54	−4.13	−9.21
Shell Thickness t = 6 mm; F = 10 kN						
Thin plate	266.1	225.6	41.1	245.5	0.445	2.19
Thick plate	259.4	220.2	40.4	238.7	0.469	2.40
Discrepancy	2.52	2.39	1.70	2.77	−5.39	−9.59
Shell Thickness t = 7 mm; F = 20 kN						
Thin plate	388.9	323.5	68.8	355.6	0.570	1.50
Thick plate	378.4	314.9	67.7	345.1	0.609	1.65
Discrepancy	2.70	2.66	1.60	2.95	−6.84	−10.00
Shell Thickness t = 8 mm; F = 20 kN						
Thin plate	295.8	245.8	58.8	268.6	0.391	1.98
Thick plate	287.8	237.8	58.1	260.6	0.423	2.16
Discrepancy	2.70	3.25	1.19	2.98	−8.18	−9.09

Table 7.5
Comparison of the Results of Both Geometrical Models

Used Formulation	Maximal von Mises Stresses (MPa)				Displacement (mm)	FoS
	Top	Bottom	Membrane	Bending		
Shell Thickness t = 3 mm; F = 5 kN						
Surface.sldprt	519.6	473.9	46.0	494.8	1.679	1.20
Sheet_Metal.sldprt	631.2	404.6	126.3	512.0	1.685	0.85
Shell Thickness t = 4 mm; F = 10 kN						
Surface.sldprt	585.08	520.0	62.8	550.5	1.462	1.06
Sheet_Metal.sldprt	744.2	451.0	179.0	574.0	1.436	0.74
Shell Thickness t = 5 mm; F = 10 kN						
Surface.sldprt	375.0	325.2	49.1	348.5	0.781	1.66
Sheet_Metal.sldprt	530.4	280.1	145.1	389.9	0.748	1.11
Shell Thickness t = 6 mm; F = 10 kN						
Surface.sldprt	259.4	220.2	40.4	238.7	0.469	2.40
Sheet_Metal.sldprt	384.4	189.5	115.9	272.4	0.443	1.52

In this section, we compared 3D FEA to 2D FEA. We discussed the impact of the adopted calculating formulation by comparing thin-plate results to thick-plate results.

We learned the following:

- What the advantages are of 2D FEA, when shell objects are studied.
- Will it be crucial to our analysis if we do not know which formulation (thin or thick shell) to adopt?
- There is no general answer to the question: Which of the two tools (Surface or Sheet Metal) for development of shell geometry is better and should be preferred? The general answer depends on the analysed object and on the user experience.

CHAPTER 8

STATIC ANALYSIS OF A FRAME BODY

8.1 BEAMS OR TRUSSES?

There are many planar or spatial frames that are better modelled by 1D finite elements (FEs) instead of the known solid or shell elements.

There are two main groups of 1D FEs used by SW Simulation.

The structure in Figure 8.1a is built of 1D elements connected to each other through joints, which are designed in a way that no moments develop in them. We are not interested how these joints are implemented in the real structure – through welding, through bold connection or through another way. More important is that this structure is exposed only to forces applied at the joints. Thus, each member of the structure is either tensed or compressed. The only inner force that develops in the member is the axial force. This axial force is constant along its length and generates an axial stress that is uniform throughout the cross section. The used structural 1D members are known as **trusses**. Trusses are commonly used in structural applications such as bridges, roofs, etc.

In finite element analysis (FEA), truss members are modelled as truss elements. The truss is a special beam element that can resist only axial deformation. It is defined by two nodes at its ends. Each node has 3 orthogonal translational degrees of freedom (DoFs).

A particular case of a planar truss FE is given in Figure 8.2a. It is pinned at the left node, and an axial force P is applied at the right node. Axial direction is along the length of the truss. The axial stress is $\sigma_x = P/A$, and the axial displacement of the right node is $u_x = PL/EA$, where P is the axial force along the length of the truss element, A is the cross-sectional area of the truss, L is the length of the truss and E is the modulus of elasticity. If we write this equation as $u_x = P.1/(EA/L) = P/K$, $K = EA/L$ is called the axial stiffness of the element and describes its ability to resist axial loads. Thus, the truss is considered analogous to an axial spring.

Entirely different is the structure in Figure 8.1b. It is exposed to a horizontal force applied at the upper left corner of the frame. If all frame members are trusses, there

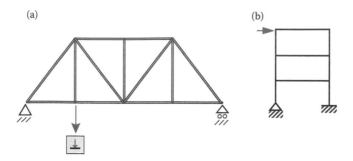

Figure 8.1
Frame structures (SW Simulation on-line help). (a) Frame structure of trusses; (b) frame structure of beams.

is no way to transfer the force to the ground, and the frame will fall aside. This does not happen because the top horizontal member is fixed to the two vertical ones, that is, the connection between them is not a joint. It can be either a fixed connection, which is to be calculated as a rigid connection, or an elastic connection, and the real stiffness of the connected members is to be considered throughout the calculation. To achieve this, 1D FEs of another type are used. These FEs are known as **beams**. Beams resist bending, shear and torsional loads. To calculate the displacements, deformations and stresses inside beams, FE software requires defining the exact cross section. The stresses vary within the plane of the cross section and along the beam. In a general case, each node of the beam element has 3 orthogonal translational DOFs and 3 rotational DOFs. Therefore, not only forces but also moments can be acting on the beam.

Figure 8.2b shows a small segment along a beam element subjected to simplified 2D inner forces – axial force P, shearing force V and bending moment M. The stresses throughout the cross section of the beam are equal to

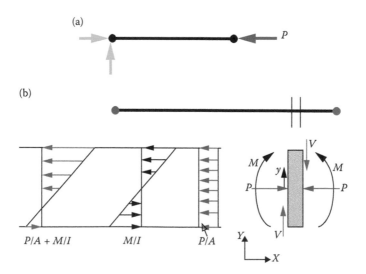

Figure 8.2
1D FEs used by SW Simulation (SW Simulation on-line help). (a) Truss FE; (b) beam FE.

- $\tau_x = V/A$ – uniform shearing stress.
- $\sigma'_x = P/A$ – uniform axial stress, similar to those of the truss elements.
- $\sigma''_x = (M/I)^*y$ – non-uniform bending stress. It is caused by the bending moment M and varies linearly with the vertical distance y from the neutral axis. The bending stress is the largest at the extreme fibres. I is the cross-sectional moment of inertia about the neutral axis in the given equation.

Each frame problem can be simplified if the frame members are modelled with 1D FEs. This significantly reduces the required system resources. The beam formulation is acceptable if the length of the beam is at least 10 times larger than the largest dimension of its cross section. Then the software starts a procedure allowing the user to choose how to model the component – as a solid body (using 3D FEs) or as a beam (using 1D FEs). Further, the user chooses the type of the 1D FEs – a beam or a truss, in relation to the characteristics of fixtures and the connections at each member end, etc.

Beam members can be straight or curved. Each straight structural member is defined by a straight line connecting two joints at its ends. Straight beam members can have a constant cross section (Figure 8.3a) or a varying cross-sectional size along their lengths (tapered beams, Figure 8.3b). Curved structural members are modelled with a number of straight beams. The cross section of each beam is assumed to be constant throughout its length.

The program meshes each member by creating a number of beam FEs. Each beam FE is defined by two end nodes and a cross section. When viewing the mesh and the results, the beam elements are represented by cylinders regardless of their actual cross section. Each structural member can be defined as a beam (first stage) and then subdivided into a number of beam FEs represented by cylinders (second stage, Figure 8.4).

(a)   (b)

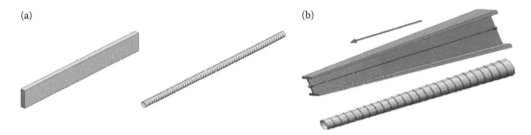

Figure 8.3
Beam members with constant or variable cross section (SW Simulation on-line help). (a) Beam member with constant cross section; (b) tapered beam member.

Figure 8.4
Modelling a structural member by beam FEs (SW Simulation on-line help).

The section discussed the difference between the truss and the beam members of the frame structures; answered the question when the simplification of the structure by adopting 1D FEs provides acceptable results; compared both types of beam members, beams with constant cross section and tapered beams; and explained how the software models them using FEs.

We outlined the difference between the truss and the beam. We obtained a criterion for the accuracy of the results when the beam model is adopted. We studied how to calculate the stresses, adopting a truss or a beam formulation. We defined the axial stiffness of a truss. We learned how the software meshes straight and curved beams, beams with a constant cross section and tapered beams.

8.2   DEVELOPMENT OF A CAD MODEL OF A 3D FRAME

The CAD model will be developed using the **Weldments** tool. After starting a new part file (File → New →Part → OK) and setting the used units "**millimeter-gram-second**" (Tools → Options → Document Properties → Units → Unit system MMGS → OK), the model will be saved as **Frame_1. sldprt**. The next stages include

1. Drawing **Sketch1** in **Front Plane** (Figure 8.5a).
2. Drawing **Sketch2** in **Right Plane** (Figure 8.5b). The sketch is perpendicular to **Sketch1**.

   The isometric view of both sketches is shown in Figure 8.5c. These sketches form the frame of the created spatial 'structural member' structure. The sketch segments define the path of the structural members. The sketched frame of the structure can be a planar sketch, a spatial one or a combination of a few sketches. It can include linear or curved entities. The sketch in the studied case is a combination of two perpendicular planar sketches.

   The next stage is defining the structural members.

   The easiest way to define a structural member is to use the **Weldments** tool.

(a)   (b)   (c)

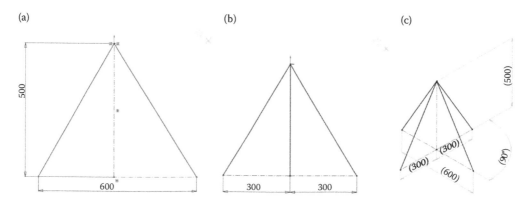

Figure 8.5
CAD model of the spatial frame sketch. (a) Sketch1; (b) Sketch2; (c) isometric view of both sketches.

3. Starting the **Weldments** tool.

To do so, right click on the **SW toolbar** and select the **Weldments** command from the pop-down menu (Figure 8.6a). The icons of some of the **Weldment** commands, being used further, are given in Figure 8.6b.

4. Adding and defining the structural members along the sketch segments of **Sketch1**.

Weldments → Structural Member ( ) → OK

There are two sub-windows, which have options that help us in defining the properties of each structural member.

The first sub-window is **Selections**, where we select the type of the profile. The software has a rich database of different profiles in ISO, ANSI INCH or custom standard. The experienced users can add self-defined cross sections of the profiles to that library. In the pop-down **Type** menu, we can select the type of the profile: rectangular, pipe, etc. (Figure 8.7c) and to continue with selecting the size of the profile in the **Size** window. There is an option **Merge arc segment bodies**, which is available only for curved entities and adjusts the arc segment to the bodies in the structural member. We have selected an ISO 30 × 30 × 2.6 square tube profile (Figure 8.7d). This means that an ISO

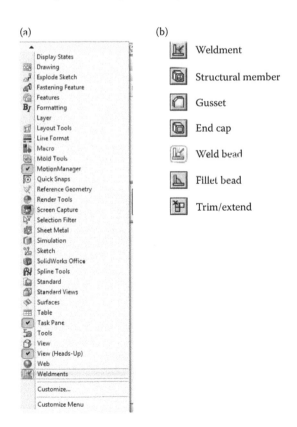

Figure 8.6
Starting Weldments tool. (a) Pop-down menu where you can start Weldments tool; (b) commands in Weldments toolbar.

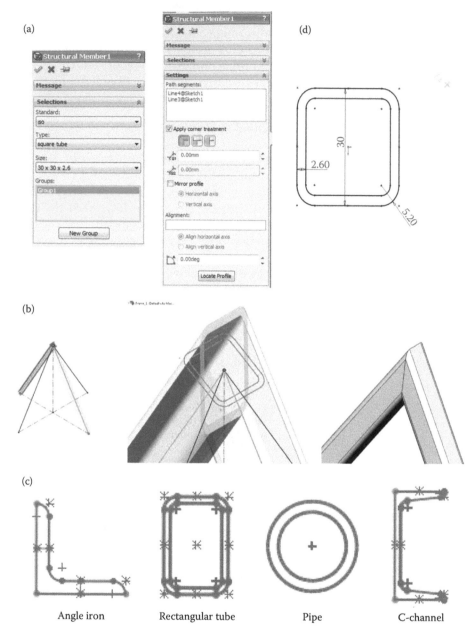

Figure 8.7

Definition of the structural members from Group1. (a) Structural member property manager – Group1; (b) selected sketch segments, cross section of the chosen profile and outer view of the defined group of structural members; (c) samples of the different profiles in the software library (SW Simulation on-line help); (d) sketch of the selected profile sample – square tube 30 × 30 × 2.6.

standard, square tube profile with outer dimension of 30 mm and a thickness of 2.6 mm, is applied to the frame. This structural profile is attached to **Group1**, whose members are selected by picking at the **Graphics area** and whose signatures are in **Path segments** in the **Settings** sub-window (Figure 8.7a). **Group1** unites the structural members, whose paths correspond to the segments of **Sketch1** (Figure 8.7b).

The second sub-window is titled **Settings** and defines the properties of the defined group of structural members (Figure 8.7a). By either clicking directly on the model components in the **Graphics area** or by picking the sketches from the **Design manager** tree, we define the segments in the group, whose signatures are listed directly in the **Path segments** window (see blue lines in Figure 8.7b). The next choice concerns either applying or not applying corner treatment. The **Apply corner treatment** option is available only for contiguous groups and defines how to trim group segments at corner intersections. This option can be modified within the study. As the **Apply corner treatment** is selected, we can choose among **End Miter** (), **End Butt1** () or **End Butt2** (). The weld gap at the corners of the segments is input in the **Gap between Connected Segments in Same Group** window (). The weld gap between the end segments of the defined group and the segments of other groups is introduced in the **Gap between Different Group Segments** window (). The next few options can also be selected:

- **Mirror Profile**, which flips the profile of the group about its **Horizontal Axis** or **Vertical Axis**
- **Alignment**, which aligns the axis of the group profile to any selected edge, construction line, etc. picked at the **Graphics area**
- **Rotation Angle** (), which rotates the structural member by a set number of degrees
- **Locate Profile** option, which zooms to the profile, so that the user can change its **pierce point**

The pierce point defines the location of the profile, relative to the sketch segment used to create the structural member. The default pierce point is the sketch origin in the profile library feature part. However, any vertex or sketch point specified in the profile can also used as a **pierce point** (Figure 8.8).

For this case study, the following options of the **Settings** are selected: the lines, which form the group; End Miter corner treatment; **Gap between Connected Segments in Same Group** is zero; No **Gap between Different Group Segments** is defined for there is only one group of structural members; no need to flip the profile for it is symmetrical – **Mirror profile** unchecked;

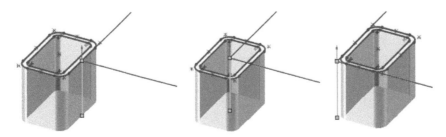

Figure 8.8
Profile sample, with three different pierce points (SW Simulation on-line help).

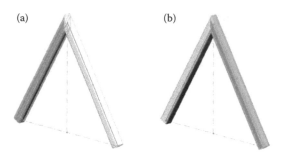

Figure 8.9
Structural members attached to Sketch1. (a) Structural Member1 [1]; (b) Structural Member1 [2].

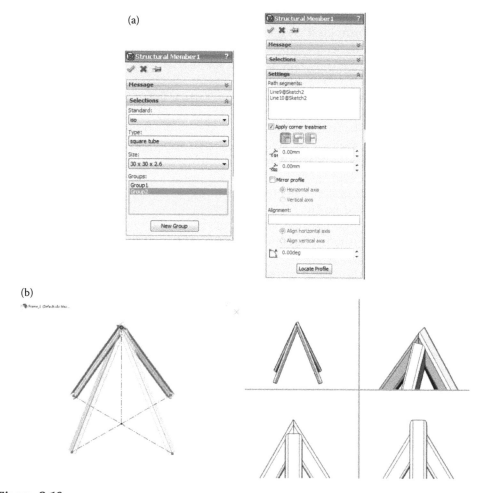

Figure 8.10
Definition of the structural members of Group2. (a) Structural member property manager – Group2; (b) selected sketch segments and outer view of both defined groups of structural members.

Rotation angle is zero and the pierce point of the profile coincides with its axis (Figure 8.7b).

After clicking **OK** and closing the **Structural member** property manager, some new lines appear in the **Design manager tree**. They include **Structural Member1** [1] (Figure 8.9a) and **Structural Member1** [2] (Figure 8.9b).

5. Adding and defining the structural members along the sketch segments of **Sketch2**.

Without closing the **Structural member** property manager, we continue with the definition of the second group of structural members. It unites the structural members, whose paths correspond to the line segments of **Sketch2**. To start this operation, we click on the **New Group** button in the **Selections** sub-window and pick the lines at the **Graphics area** (Figure 8.10). The user can see how the software forms the upper connection of the structure and, if necessary, how to modify it through **Apply corner treatment** options.

In this section, the user develops a CAD model of a 3D frame using the **Weldments** tool and by defining structural members along the line segments of two perpendicular sketches.

We learned how to create a CAD model of a 3D frame using a combination of two perpendicular sketches. The path of each structural element corresponds to the line segments of these sketches. We defined the cross section of the elements using an existing library of structural elements. We explained the different ways of corner treatment and the required input data.

8.3 CALCULATION OF A 3D FRAME OF TRUSSES

8.3.1 Pre-Processor and Processor Stages

Our next task is to perform a static analysis of the developed 3D frame using truss elements. Therefore, we have to define immovable fixtures at the end of each structural member and to 'tell' the software that the connection at the top of the structure is a joint. Consequently, the external loads are limited to forces at the joints.

We start a static analysis. Its title is **Static_Truss** to remind us that we will use truss elements. The material of all bodies is **Aluminum Alloy 1060**, with the following properties: elastic modulus – 69 GPa; Poisson's ratio – 0.33; shear modulus – 27 GPa; mass density – 2700 kg/m^3; tensile strength – 68.94 MPa; yield strength – 27.57 MPa. As we intend to use truss elements, only elastic modulus and yield strength are required for performing the calculations. As gravitational loads will be considered, the density is also required.

After right click on the name of each body (Structural Member), a pop-up menu appears (Figure 8.11a), where we can select one of the two options – either **Treat as Solid** or **Edit definition**. The icon in front of the name of the body (🢂) prompts that the structural member is modelled as 1D body, that is, either truss or beam FEs will be

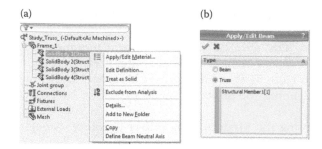

Figure 8.11

Definition of the truss structural members. (a) Simulation analysis pop-down menu; (b) apply/ edit Beam property manager – Truss checked.

used throughout the structural analysis. If we pick the **Treat as Solid** command, the software automatically starts to consider the body as a solid one and the icon in front of its name changes (🗐). Further, this body will be meshed with 3D solid FEs. Truss FEs are preferred for this example. Thus, we pick the **Edit definition** command, and the **Apply/Edit Beam** property manager appears. This property manager allows the user to define a beam or a truss. Our choice directly influences the transfer of forces and deformations. We pick **Truss** (Figure 8.11b) and then **OK**. The icon in front of the body's name automatically changes to ⬣.

When defining beams or trusses, the software automatically generates a new group in the **Simulation Analysis** tree. This new group is titled **Joint group** (⬣) and calculates the joints in the analysed structure. The software identifies a joint at the free end of a structural member or at the intersection of two or more structural members. The joint coincides with the pierce point of the weldment profile. Hence, it is recommended to locate the **pierce point** (Figure 8.8) at the centre of gravity of the weldment profile (Figure 8.12). If so, the axial loads generate axial stresses only.

After right clicking on **Joint group** and selecting **Edit** from the pop-down menu, the **Edit Joints** property manager opens (Figure 8.13a). It combines three sub-windows: **Selected Beams**, **Results** and **Criteria**. There are two options in the **Selected Beams** sub-window: **All** – when the software considers all structural members when calculating the joints; and **Select** – when the software considers only the selected structural members (⬣) when calculating joints. Clicking the button **Calculate** calculates the joints in the structure. As a result, the software colours in pink (⬤) the joints where two or more structural members intersect and colours in green (⬤) the joints at the ends of the structural members, as well as their mid-points (Figure 8.13b).

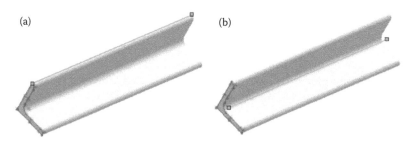

Figure 8.12

Definition of the joints for truss structural members (SW Simulation on-line help). (a) A pierce point at the vertex of the angle profile; (b) a pierce point at the centre of gravity of the angle profile.

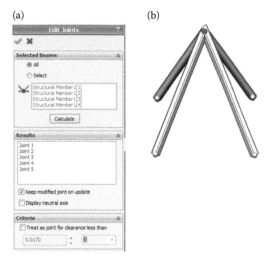

Figure 8.13
Calculation of the joints. (a) Edit Joints property manager; (b) calculated joints.

The list of all calculated joints is automatically displayed in the **Results** sub-window. By clicking on a joint from the list, the software highlights it, and we can edit its properties or even delete it. Selecting the **Keep modified joint on update** allows the user to save the modifications in the model.

The **Criteria** sub-window enables definition of joints between non-touching structural members within a certain distance (tolerance). The software suggests an optimal tolerance value, but the user can overwrite it by selecting **Treat as joint for clearance less than**.

Restraints can be applied to joints only. Only translational restraints can be applied to truss joints. There are 3 translational degrees of freedom at each node (joint). Consequently, **Fixed Geometry** (🔧) and **Immovable (No translations)** (🔧) restraints act similarly at truss joints as no rotations are considered. Zero or non-zero prescribed translations can also be applied. The joints at which restraints are applied are selected by picking at the **Graphics area** and are listed in the joint window (⚹ , Figure 8.14).

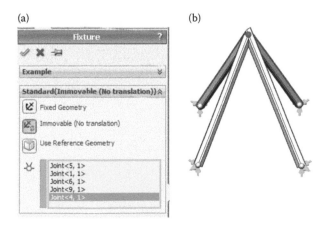

Figure 8.14
Application of the fixtures. (a) Fixture property manager for structures of 1D members; (b) 3D structure with applied fixtures.

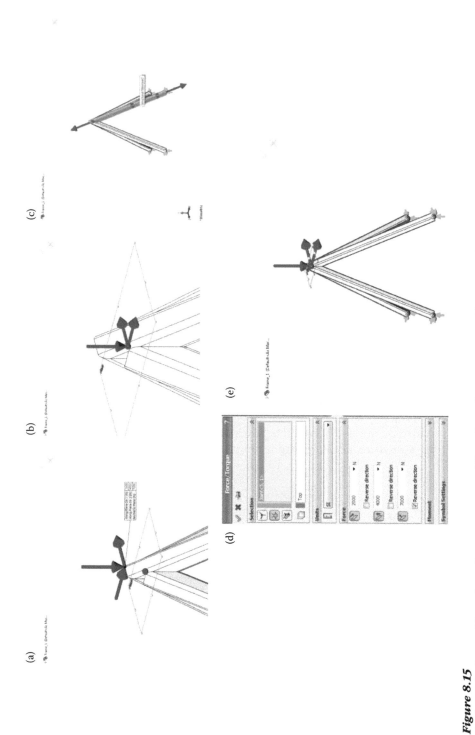

Figure 8.15

Applying the forces. (a) Applying forces at a vertex; (b) applying forces at a joint; (c) applying forces at a structural member; (d) Force property manager; (e) applied forces.

The software applies only concentrated forces at joints or at reference points. The software automatically ignores any forces applied normally to the truss. Only axial forces generated in the elements are considered. After activation of the **Force/Torque** property manager, only force options are accessible. Forces can be applied at **Vertices** or **Points** (⊤, Figure 8.15a), at **Joints** (⊠, Figure 8.15b) or at **Structural members** (⬚). When applying a force at a structural member, it can be either a tensile or a compressive force along the axis of the member (Figure 8.15c). We apply the force at the top joint (Figure 8.15d and e). When applying the force at a vertex or a point, or at a joint, a reference entity **Face, Edge or Plane for direction** (⬚) must be selected. The values of the force components are related to that selection. We have selected Top Plane as a reference plane (Figure 8.15d and e). The selected unit for force components is newton (N). The values of the components are as follows: **Along Plane Dir 1** (⬚) – 2000 N; **Along Plane Dir 2** (⬚) – 4000 N; and **Normal to Plane** (⬚) – 7000 N with **Reverse direction** selected (Figure 8.15d).

Regarding the meshing, there are no options in meshing trusses. A straight truss is represented by one truss element. The axial stress is constant throughout the cross section and along the truss. The variation of the axial deformation is linear. Meshed truss members are displayed as solid cylinders regardless of their actual cross-sectional shape. The meshed structure consists of four FEs and five nodes, which coincide with the joints (Figure 8.16).

8.3.2 Viewing the Results

The use of truss FEs provides results for axial stresses and forces, displacements and deformed shape plots. Forces and stresses of a truss member are constant throughout the cross section and along the truss. In the stress plot, each truss element appears in one colour (Figure 8.17a).

Inner forces are displayed by activating the **Beam Diagrams** property manager after right clicking the **Results** menu. We can pick some or all trusses (⬚). Automatically the software suggests all structural members to be selected. We select the **component** (⬚) and the **units** (⬚) of the displayed inner force – axial force, measured in newtons. The **Beam Diagram Width** option controls the width of beam plots for better visibility and is set to 30 (Figure 8.17c). The force in a truss member equals the axial stress multiplied by its cross-sectional area (Figure 8.17b).

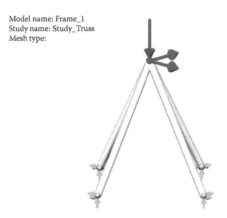

Model name: Frame_1
Study name: Study_Truss
Mesh type:

Figure 8.16
Meshed structure.

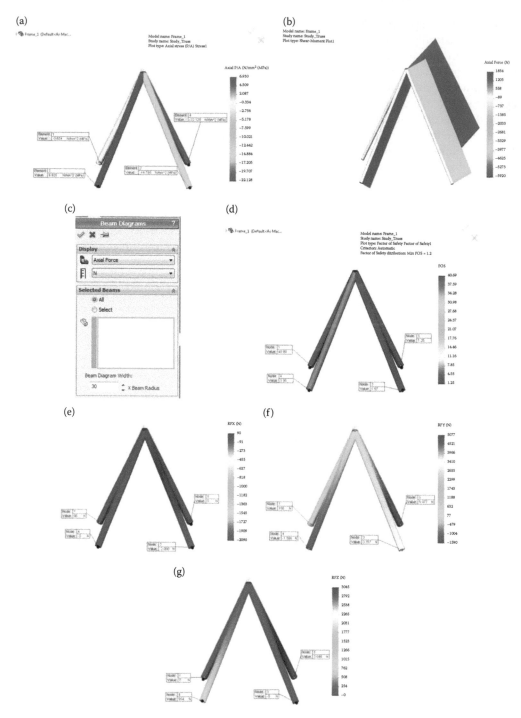

Figure 8.17
Results of truss structure calculation. (a) Axial stresses; (b) axial force diagram; (c) Beam Diagrams property manager; (d) FoS plot; (e) RX reaction in each truss; (f) RY reaction in each truss; (g) RZ reaction in each truss.

Table 8.1
Results of Truss Structure Analysis

Results	Structural Member 1	Structural Member 2	Structural Member 3	Structural Member 4
Stresses (MPa)	−0.675	−14.780	6.930	−22.128
Axial forces (N)	−180	−3955	1854	−5920
FoS	40.89	1.87	3.98	1.25
Reactions				
RX (N)	90	−2090	0	0
RY (N)	156	3357	−1590	5077
RZ (N)	0	0	954	3046
Resultant reaction force (N)	180	3955	1854	5920

The displacements vary linearly between the ends. The displacements at the top node are UX = 0.118 mm; UY = −0.075 mm; UZ = −0.238 mm; and UREZ = 0.277 mm.

The reactions are plotted in Figure 8.17e–g. The values are listed in Table 8.1. Axial forces and stresses at each truss can be listed and saved as a *.csv or *.txt file.

In this section, we made a FE analysis of a spatial frame using truss FEs.

We learned

- How to transform structural members in trusses
- What the restrictions are regarding the application of external loads and fixtures
- How the software visualizes truss members after meshing the frame
- The main results provided by the program

8.4 CALCULATION OF A 3D FRAME OF BEAMS

8.4.1 Pre-Processor and Processor Stages

When the ratio of a body's length over the largest orthogonal cross-sectional distance from the centroid is larger than 3.0, the software offers the option **Treat as Beam** in the pop-up menu that is displayed after right clicking on the name of the part. If this option is selected, the icon in front of the name of the part changes to **Beam** (≷). By default, the software chooses this option for all structural members. The beams can have a constant cross section or a varying cross-sectional size along their lengths. The second group of beams is known as tapered beams (≷).

Before defining the beam properties, the material of the structure is set. This is Aluminum Alloy1060. When beam FEs are used, **the modulus of elasticity** and **Poisson's ratio** of the material are always required. **Density** is required only if gravitational loads are considered. Modulus of elasticity and Poisson's ratio are 69,000 MPa and 0.3, respectively.

The properties of the beam are introduced through the **Apply/Edit Beam** property manager. At first, the user defines whether the member will be treated as a truss or as a beam. Differing from the truss, the beam resists axial, bending and torsional loads. If the beam option is picked, the user controls the transfer of forces and moments at each end. Thus, any of the force or moment components at the member's ends can be set to zero. The input restraints apply to joints and hence to all beams' ends that meet at those joints also. The options specified by the **Apply/Edit Beam** property manager override restraints. For example, if you define a beam end as a **Hinge** and apply a **Fixed** restraint to the associated joint, the specified beam end acts as a **Hinge** and does not carry any moment.

The **Apply/Edit Beam** property manager combines four sub-windows (Figure 8.18a).

The first sub-window is **Type** and it enables the user to select the type of the picked member – a truss or a beam (Figure 8.18b). The picked member is visualized at the **Graphics area** (Figure 8.18e).

The next two sub-windows introduce the restraints at each end of the beam. End 1 is coloured in red (●), while End 2 is coloured in blue (●, Figure 8.18e).

The sub-window **End1 Connection** sets the forces and the moments at the first end of the beam. The **End2 Connection** sub-window has the same set of options (Figure 8.18c). The user chooses among the following options:

- **Rigid** – No forces or moments are released at this end. If restraints are applied to the associated joint, the restraint condition fully defines the transfer of forces and moments. If no restraints are applied, continuity is assumed at the associated joint. This option is used unless there is a reason to release (set to zero) force or moment components at the end.
- **Hinge** – The end can rotate freely and does not transfer any moments to the joint. This condition is applied to all beam ends meeting at the joint to define the joint as an intermediate hinge.
- **Slide** – The end can translate freely and does not transfer any forces to the joint.
- **Manual** – If this option is checked, the user must manually specify for each force or moment component whether it is known to be zero. To do so, the user either checks or not the following options:
  - Hinge, first direction – sets the moment about the first direction of the cross section to zero. The end can rotate about this direction.
  - Hinge, second direction – if checked, the moment about the second direction of the cross section is zero. The end rotates about this direction.
  - Hinge, along beam – the options must be checked if the moment about the axial direction of the beam is known to be zero. The end can rotate about this direction; hence, this beam end is not exposed to torsion.
  - Slide, first direction – if the force in the first direction of the cross section is known to be zero, the user must check this option. The end can translate freely along this direction.
  - Slide, second direction – select this option if the force in the second direction of the cross section is known to be zero. The end translates along that direction.
  - Slide, along beam – this option sets the force in the axial direction of the beam to zero. The end can translate along this direction.

The directions of the beam are shown in Figure 8.18f. The red arrow on the right picture shows the positive axial direction, the green arrow shows the positive direction 1 and the blue arrow shows the positive direction 2 for each beam member.

For this case study, Rigid option at both ends of each beam member is checked (Figure 8.18c).

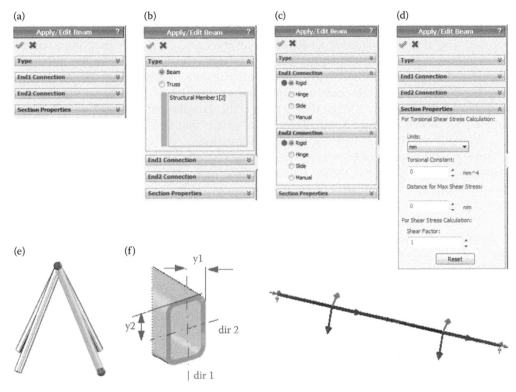

Figure 8.18
Definition of the constraints at the beam ends. (a) Apply/Edit Beam property manager; (b) Type sub-window; (c) End Connection sub-windows; (d) Section Properties sub-window; (e) view of the selected structural member; (f) directions of the beam (SW Simulation on-line help).

The fourth sub-window of the **Apply/Edit Beam** property manager is **Section Properties**.

It allows the user to select the unit of length for the calculation of torsional constant and distance for maximum torsional shear (mm in the example). Further, the user can introduce the calculated **Torsional Constant** (K). Torsional stiffness constant has a dimension of the length to the fourth power and is a function of the cross section. To calculate the torsional constant, the user can utilise the special tables (Formulas for torsional deformation and stress, Formulas for Stress and Strain of Roark and Young, for example). The next window defines the **Distance for Max Shear** (CTOR). This is the distance from the centre of the section to the point of maximum torsional shear. The distance for maximum shear depends on the cross section. The maximum torsional shear stress is then calculated from $\tau_{MAX} = (T/K) * CTOR$, where T is the applied torque. The software requires a **shear factor** (**SF**) for the shear stress calculation. This factor is the ratio of (effective area under shear/beam's cross-section area). It depends on the cross-section's shape and is less than 1. For example, for a rectangular section, $SF = \tfrac{2}{3}$, and for a solid circular section, $SF = \tfrac{3}{4}$. The maximum shear stresses in the two local directions are $V_1/A*SF$ and $V_2/A*SF$, where V_1 and V_2 are the two shear forces, A is the cross-sectional area and SF is the shear factor, as input by the user. The **Reset** button resets the variables to their default values.

After setting the type of each beam member, we have to recalculate the joints. We will do this according to the instructions in Section 8.3. It is important to remember that the program creates a node at the centre of the cross section of each joint member. Sometimes, due to trimming and the use of different cross sections for different members, the nodes of members associated with the joint may not coincide. Then, the program creates special elements near the joint to simulate a rigid connection based on geometric and material properties. Further, the **Display of neutral axis** button can also be checked (Figure 8.13a). The neutral axis is that fibre of the beam member, whose length remains constant during the deformation.

The software allows the user to identify a new neutral axis and thus to override the neutral axes of beams automatically selected by the program. To do this, the user must start the **Identify Neutral Axis** property manager by right clicking on the name of the structural member and then clicking the **Define Beam Neutral Axis** command (Figure 8.19). The user selects an edge of the beam's body that is parallel to the desired orientation of the new neutral axis (⬚). The use of that tool is worthy in cases where the identification of neutral axes by the software may not be accurate, and there is a need to modify the direction of the beam's neutral axis, for example, for short structural members with a length-to-width ratio < 3. If the edge selection is invalid, a message appears to inform that the beam's section properties are incorrect and prompts selecting a valid edge.

Performing FE analysis with beam FEs, the user can apply restraints to joints only. There are 6 degrees of freedom at each joint. Thus, zero or non-zero prescribed translations and rotations can be applied. **Fixed Geometry** fixtures will be applied at the free end of each structural member (Figure 8.20a).

Afterwards, the external loads are applied. There are no restrictions that these loads be forces and be applied only at joints, as is the case when truss FEs are used. The applied loads are as follows:

- **Gravity** (the red arrow, Figure 8.20b).
- **Torque** applied to the top joint (⊠) and equal to 100 Nm (the violet arrow, Figure 8.20c).
- **Non-uniform load**, distributed according to parabolic (◪) law. The total value of the load is 500 N (the orange arrows, Figure 8.20d). This load is applied at the beam (◩) picked in the **Selection** sub-window. When non-uniform load is input, the software does not allow selection of multiple members. Thus, the introduction of the linear non-uniform loads is done in two consequent operations – for every structural member separately.

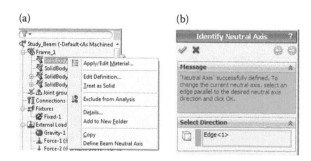

Figure 8.19
Definition of a new beam neutral axis. (a) Start of the procedure; (b) Identify Neutral Axis property manager.

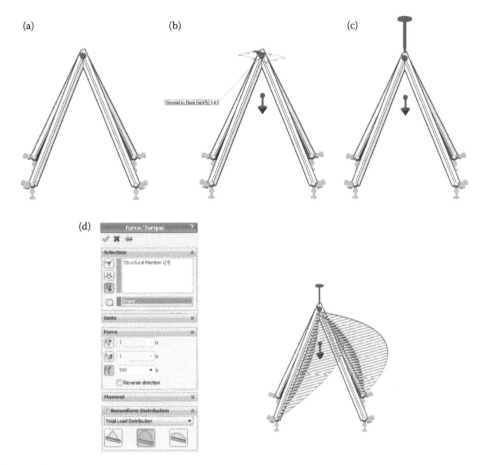

Figure 8.20
Defined the boundary conditions. (a) Fixed Geometry fixtures; (b) gravity is applied; (c) torque is applied; (d) applied non-uniform distributed loads.

Further, some additional explanations about the input of non-uniform loads are given. When a uniform load along the beam is applied, the units are set to newtons per meter and the total applied force is calculated by the software as a product of the load value and the length of the beam. For non-uniform loads, this option is not available. When a non-uniform distribution is selected, the user must choose one of the following options:

- **Total Load Distribution** – It distributes the total force/moment along the length of the beam. No loads are applied at the ends of the beam. The shape of the distribution can be triangular, parabolic or elliptical. The total value of the load equals the value introduced in the property manager (Figure 8.20d).
- **Centred Load Distribution** – Applies the force/moment at the centre of the beam. Loads decrease on either side of the centre according to the selected triangular, parabolic or elliptical distribution and are defined per unit length. No loads are applied at the ends of the beam. The value introduced in the property manager equals the centre/maximal value of the load. The total applied load is a function of the beam's length.

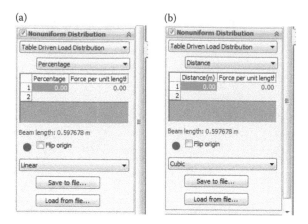

Figure 8.21
Definition of table-driven loads. (a) Table-driven percent load, using linear law interpolation for intermediate values; (b) table-driven distant load, using cubic law interpolation for intermediate values.

When **Total Load Distribution** or **Centred Load Distribution** is picked, the user selects among the following:

- **Triangular distribution** (⬛), which distributes the total or the centred load along the length of the beam in a triangular distribution.
- **Parabolic distribution** (⬛), which distributes the total or the centred load along the length of the beam in a parabolic distribution.
- **Elliptical distribution** (⬛), which distributes the total or the centred load along the length of the beam in an elliptical distribution.

The last existing option for the description of the law distribution is the **Table Driven Load Distribution** (Figure 8.21). It distributes the force values at specific locations along the length of the beam. The locations can be specified either as percentages or as distances from one end of the beam. If **Percentage** is selected, it lets entering the locations of the specified force values along the length of the beam as percentage values of the total beam length. For each percentage entry in the table, the associated force per unit length must be typed. If **Distance** is picked, the distances from the origin of the intermediate beam locations must be entered. For each distance entry in the table, the associated force per unit length must be typed. Checking the **Flip origin** reverses the tarting point of the force distribution to the opposite joint of the beam. The starting point is highlighted with an icon of a red sphere. An arrow indicates the direction of the force distribution in the **Graphics area**. The tab below guides the user in defining the interpolation scheme for intermediate beam locations not specified in the table. The user can choose between **Linear** or **Cubic** functions. The defined table-driven load distribution data can be saved to a *.csv or *.txt file by clicking **Save to file**. This file includes comma-separated values, which can be edited or viewed by a text editor or by Microsoft Excel package. The opposite command is **Load from file** (Figure 8.21).

Regardless of their actual cross-sectional shape, meshed beam members are displayed as hollow cylinders, which is the same way the truss members are displayed. The number of uniform elements is defined automatically, so the user can view the variation of deformation and stresses along the length of the member (Figure 8.22a). Differing the truss FEs, beam FEs enable the **Apply Mesh Control** (⬛, Figure 8.22b).

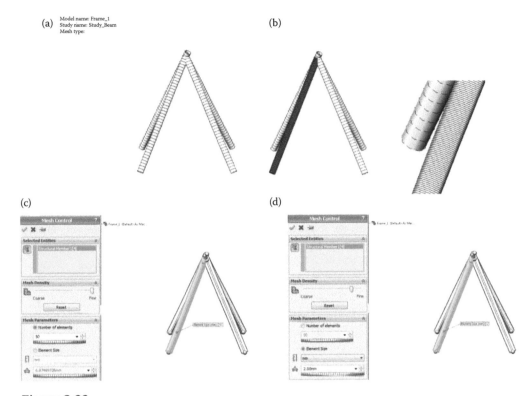

Figure 8.22

Meshing the beam structure. (a) Meshed beam structure; (b) meshed beam structure with applied mesh control at one of the members; (c) controlling the number of FEs; (d) controlling the size of FEs.

There are two ways to apply mesh control: by defining the number of FEs along the picked structural member (Figure 8.22c) or by defining their size (Figure 8.22d). Mesh control can be applied to each member separately or to a group of members.

8.4.2 Viewing the Results

Results for each element are presented in its local directions (Figures 8.18f and 8.23). There is no averaging of stresses for beam elements. The user can view uniform axial stresses, torsional and bending stresses in the two orthogonal directions (dir 1 and dir 2) and the worst stresses on extreme fibres generated by combining axial and bending stresses.

Starting the **Settings** property manager (⟦⟧), the user can select the option **Show beam direction** and thus see the local directions for every plot (Figure 8.23b). The green arrow always shows the positive **direction 1**, and the blue arrow shows the positive **direction 2**. Generally, the beam section is subjected to an axial force P and two moments $M1$ and $M2$ (Figure 8.23a). The moment $M1$ is about the axis along direction 1, and the moment $M2$ is about the axis parallel to direction 2.

The software provides the following options for viewing stresses (Figure 8.24):

- **Axial:** Uniform axial stress = P/A (Figure 8.24a).
- **Bending in local direction 1:** Bending stresses due to $M2$. This is referred to as **Bending Ms/Ss** in the plot name, title and legend (Figure 8.24c).
- **Bending in local direction 2:** Bending stress due to $M1$. This is referred to as **Bending Mt/St** in the plot name, title and legend (Figure 8.24d).

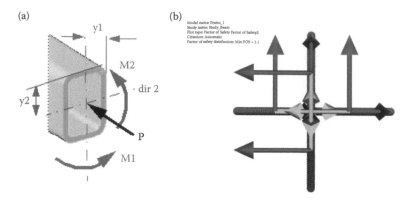

Figure 8.23
Positive directions of inner forces. (a) Positive inner forces for a beam FE (SW Simulation on-line help); (b) positive directions for the studied structure – top view.

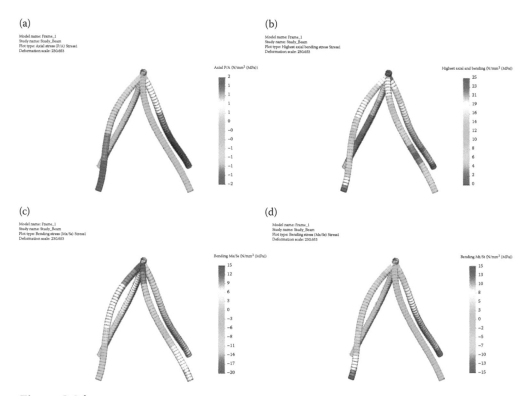

Figure 8.24
Plots of normal to the cross section of the beam member stresses. (a) Axial stresses; (b) highest axial and bending stresses; (c) bending in local direction 1 (Ms/Ss); (d) bending in local direction 2 (Mt/St).

The software automatically calculates the highest stresses at the critical point on the cross section by combining axial and bending stresses due to *M1* and *M2*. This is the worst case and the recommended stress to view (Figure 8.24b). In general, the software calculates four stress values at the extreme fibres at each end. When viewing the worst case stresses, the software shows one value for each beam segment. This value is the largest in magnitude out of the eight values calculated for the beam segment. These values are accurate for beam with cross sections that are symmetric in two directions. These values are also conservative for all other cases.

The software plots the displacements of each node, parallel to the global coordinate axes as well as the resultant displacement (Figure 8.25).

While using other types of FEs, the FoS plot is also available. Its minimal value is 1.11 (Figure 8.26).

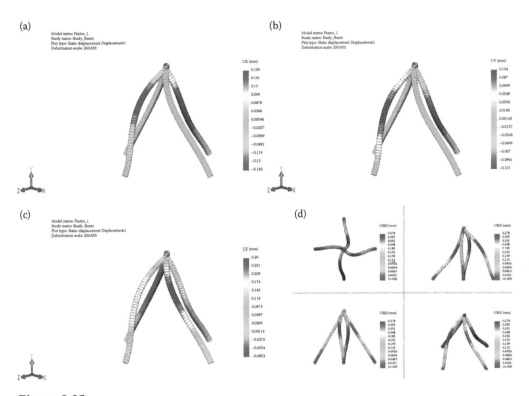

Figure 8.25
Plots of normal to the cross section of the beam member stresses. (a) Displacement of nodes parallel to X; (b) displacement of nodes parallel to Y; (c) displacement of nodes parallel to Z; (d) resultant displacement from different points of view.

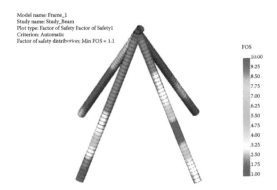

Figure 8.26
Factor of safety plot of the analysed structure.

Figure 8.27
Starting Define Beam Diagrams procedure.

Clicking **Define Beam Diagrams** from the right-click pop-up **Results** menu (Figure 8.27) allows the user to plot the inner force diagrams for structural members (Figure 8.28).

Some of the inner force values are given in Table 8.2.

The graphs of some of the inner force functions along the member axes are given in Figures 8.29 and 8.30.

8.4.3 FE Analysis, When There Are Hinge Connections at Both Ends of All Beam Members

Further, the results of the structure, when there are hinges at both ends of each beam member, are provided. To transform the model, all rigid constraints are replaced by hinge ones (Figure 8.18c) through the **Apply/Edit Beam** property manager.

The minimum FoS strongly reduces to 0.586. This is a very good example to show one of the easiest ways to make the structure more reliable by simple change of the type of the connections between structural members. If the structure with rigid constraints is safe enough, the structure with hinges at both ends is not recommended.

The inner force diagrams of the newly studied structure are shown in Figure 8.31. There is no torsion at the beams because of the newly defined beam constraints. Even more, all moments at the fixtures are zero, that is, the applied **Fixed Geometry** ($\boxed{\mathscr{E}}$) fixtures act as **Immovable** ($\mathscr{E}_0$), which means that when hinges replace the rigid constraints, the software automatically ignores the fixing moments and replaces the **Fixed Geometry** fixtures with **Immovable** fixtures.

Table 8.3 provides the values of the inner forces as well as the reaction forces at each fixture.

The graphs of the polynomial inner force functions are shown in Figure 8.32. As you can see, there is zero bending moment in direction 2 at both ends of each beam.

In this section, we made a FE analysis of a spatial frame using beam FEs. We applied different external loads at the analysed structure: gravity, torque and non-uniform parabolic linear load. We varied the restraints at the ends of beam members and compared the results.

We learned how

- To transform 1D members of the structure in beam members and how to define the restraints at beam ends
- To apply non-uniform linear loads at beam members
- To view inner force diagram and stresses as well as the reaction components
- Different end constraints influence the final results

We showed how a simple modification of a structural members' connection can increase the reliability and the safety of the structure, without adding new components or significant cost increase.

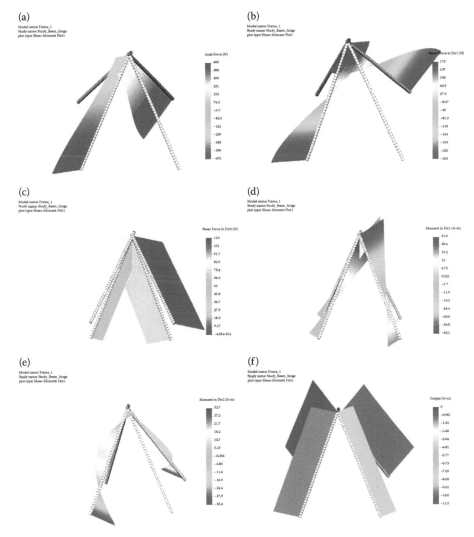

Figure 8.28
Inner force diagrams. (a) Axial force diagram; (b) diagram of shear force in Dir 1; (c) diagram of shear force in Dir 2; (d) diagram of moment in Dir 1; (e) diagram of moment in Dir 2; (f) diagram of torque.

Table 8.2
Results for the Analysed Frame

| Structural Member | Inner Force | | | |
	Shear Force in Dir 2 (N)	Minimum Moment in Dir 1 (N m)	Maximum Moment in Dir 1 (N m)	Torque (N m)
1	43.76	−8.1	16.6	−11.5
2	110.03	−43.05	21.8	−8.4
3	87.61	−32.4	16.9	−10.4
4	82.06	−15.86	31.65	−10.0
Reactions				
	RX (N)	RY (N)	RZ (N)	Total Reaction (N)
1	5.08	6.51	43.8	44.5
2	0.473	5.31	−110	110
3	82.1	272	−469	548
4	−87.6	−267	−461	540
	MX (N m)	MY (N m)	MZ (N m)	Total Moment (N m)
1	12.8	5.94	0.316	14.1
2	−22.9	−4.37	0.522	23.3
3	−33.4	0.391	−18.7	38.3
4	−32.7	0.256	19.9	38.2

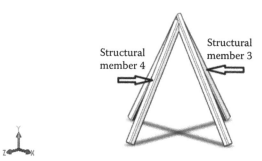

Figure 8.29
Structural members 3 and 4 of the analysed frame.

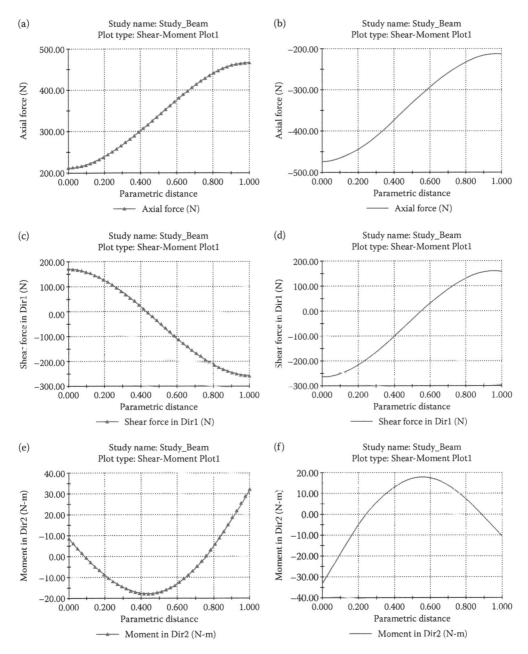

Figure 8.30
Inner force graphs of structural members 3 and 4. (a) N force for structural member 3; (b) N force for structural member 4; (c) shear force in Dir 1 for structural member 3; (d) shear force in Dir 1 for structural member 4; (e) moment in Dir 2 for structural member 3; (f) moment in Dir 2 for structural member 4.

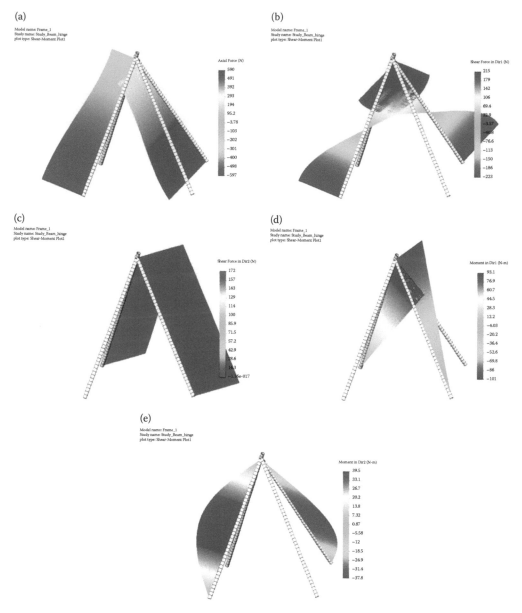

Figure 8.31

Inner force diagrams of the structure with hinged members. (a) Axial force diagram; (b) diagram of shear force in Dir 1; (c) diagram of shear force in Dir 2; (d) diagram of moment in Dir 1; (e) diagram of moment in Dir 2.

Table 8.3
Results for Analysed Frame while Rigid Restraints Are Replaced by Hinges

Structural Member	Shear Force in Dir 2 (N)	Inner Force		Torque (N m)
		Minimum Moment in Dir 1 (N m)	Maximum Moment in Dir 1 (N m)	
1	161.22	0	90.9	0
2	171.7	−101.1	0	0
3	0	0	0	0
4	0	0	0	0
Reactions				
	RX (N)	RY (N)	RZ (N)	Total Reaction (N)
1	3.2	7.05	161	161
2	−3.2	6.73	−172	172
3	0	398	−498	637
4	0	−395	−488	628

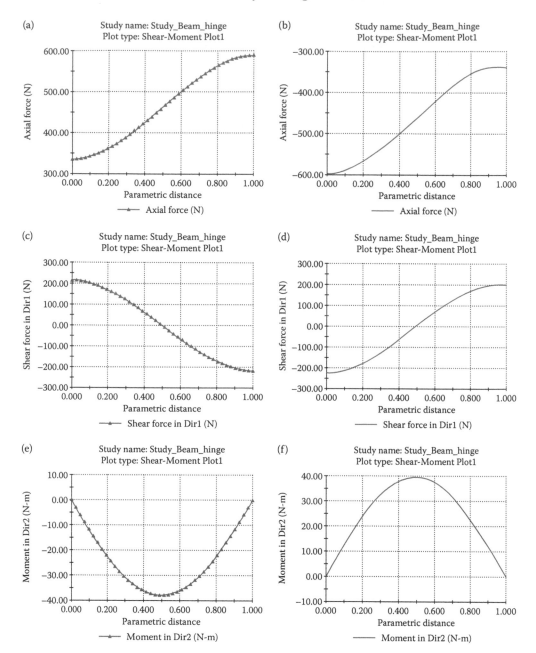

Figure 8.32
Inner force graphs of structural members 3 and 4, when there are hinge constraints at beam ends. (a) N force for structural member 3; (b) N force for structural member 4; (c) shear force in Dir 1 for structural member 3; (d) shear force in Dir 1 for structural member 4; (e) moment in Dir 2 for structural member 3; (f) moment in Dir 2 for structural member 4.

CHAPTER 9

STATIC ANALYSIS OF A COMPLEX STRUCTURE

9.1 CAD MODEL OF THE STUDIED STRUCTURE

The analysed structure unites beams, shell and a solid body (Figure 9.1).

The type of the model is 'part' (File → New → Part → OK), yet some additional contact conditions will be added to the FE model. The used unit is '**millimetre-gram-second**' (Tools → Options → Document Properties → Units → Unit system MMGS → OK). The model is saved as **Frame_2. sldprt**.

To develop the structure, the sketches and the structural members from the previous frame are updated. To ease the development of the CAD model, some of the instructions are repeated and the differences in establishment of the components are emphasised. We will start with the following:

1. Drawing the **Sketch1** in the **Top** plane (Figure 9.2a).

Sketch → Sketch1 → OK

2. Establishment of the two horizontal structural members

Weldments → Structural Member (▣) → OK

We have chosen to use ISO angle iron profiles 35 × 35 × 5, mirrored and aligned as shown in Figure 9.2b–d.

3. Definition of **Plane2**, which is perpendicular and symmetric to the drawn sketch:

Features → Reference Geometry → Plane (◇) → OK

To define the plane, both edges, where the two angle iron profiles connect each other, are picked (Figure 9.3a and b).

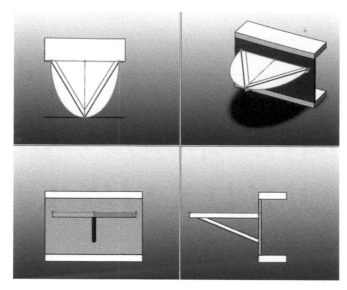

Figure 9.1
Different 3D views of the designed structure.

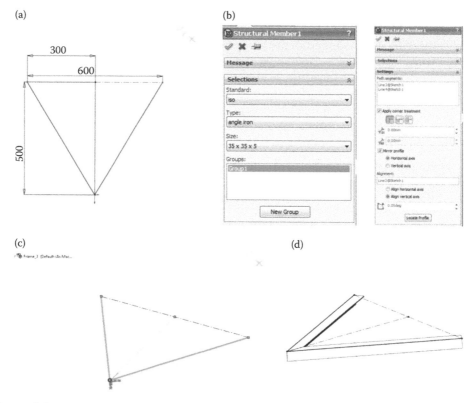

Figure 9.2
Development of CAD model of the structure – stages 1 and 2. (a) Sketch1 in Top plane. (b) Selecting the members and disposing the selected profiles of Group1. (c) Graphic area view when Group1 is introduced. (d) Establishment of both structural members.

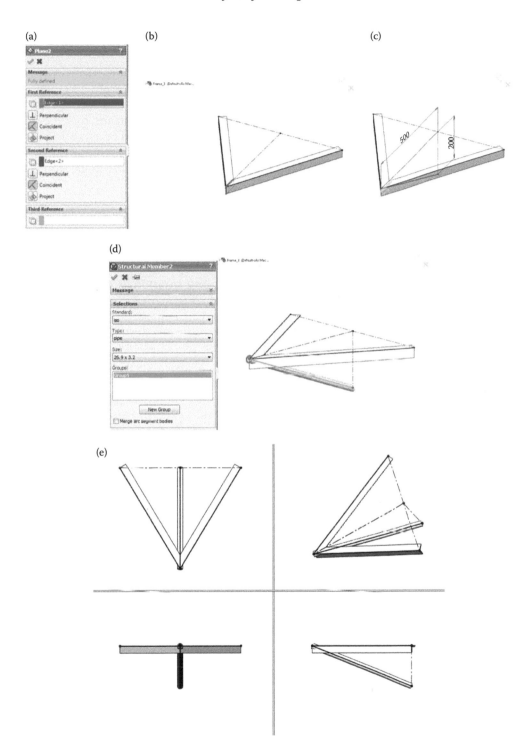

Figure 9.3
Development of CAD model of the structure – stages 3, 4 and 5. (a) Plane property manager. (b) Graphic area view when Plane2 property manager is open. (c) Sketch2. (d) Definition of Structural Member2. (e) Structural members of the two groups.

4. Drawing of **Sketch2** in the newly created plane. This is a triangle, whose dimensions are given in Figure 9.3c.
5. Definition of the third structural member (Figure 9.3d and e):

Weldments → Structural Member (⬚) → OK

We have chosen to use ISO pipe profile 26.9 × 3.2. This means that the outer diameter of the pipe is equal to 26.9 mm and the thickness of its walls is 3.2 mm.

6. Trimming the interfering parts of the pipe at its connection to the angle iron profiles:

Weldments → Trim/Extend (⬚) → OK

We will cut the upper end of the pipe profile (coloured in yellow, Figure 9.4b and c) by picking it in the **Bodies to be Trimmed** sub-window (Figure 9.4a), selecting the **Bodies** option in the **Trimming boundary** sub-window (Figure 9.4a) and picking the two angle iron profiles (coloured in pink, Figure 9.4b and c). The ready structural members' connection is shown in Figure 9.4d.

7. Definition of **Plane3**.

Features → Reference Geometry → Plane (⬚) → OK

This plane is defined by two crossing constructive lines of **Sketch1** and **Sketch2**. We can see them, coloured in pink and violet, in Figure 9.4f. The colours correspond to the colours of the windows of the **First Reference** and **Second Reference** sub-windows (Figure 9.4e). The newly defined vertical blue plane is shown in Figure 9.4f.

8. Drawing of **Sketch3** in **Plane3** (Figure 9.4g). This sketch outlines the supporting plate.
9. Establishment of the supporting C-channel part.

   The first stage includes the extrusion to 18 mm of the middle area of **Sketch3** (Figure 9.5a and b).

Features → Boss/Extrude (⬚) → OK

The second stage includes the extrusion to 200 mm of the side areas of **Sketch3** (Figure 9.5c and d). Thus, the C-channel form is made (Figure 9.5e).

10. Trimming the profiles.

    As you see, parts of the profiles interfere with the supporting plate (Figure 9.6a). Therefore, it is necessary to trim and discard these segments of the profiles. To do so, we use

Weldments → Trim/Extend (⬚) → OK

All **Bodies to be Trimmed** (the structural members) are picked and coloured in yellow, whereas the **Trimming Boundary** (trimming face) is

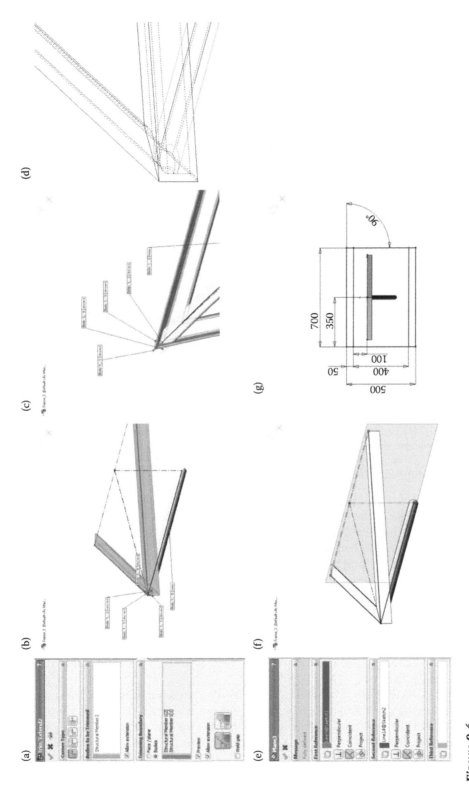

Figure 9.4

Development of CAD model of the structure – stages 6, 7 and 8. (a) Trim/Extend property manager. (b) Graphic area view. (c) Graphic area view – detail. (d) The upper pipe connection. (e) Plane3 property manager. (f) Defined Plane3. (g) Sketch3 in Plane3.

(a)   (b)   (c)

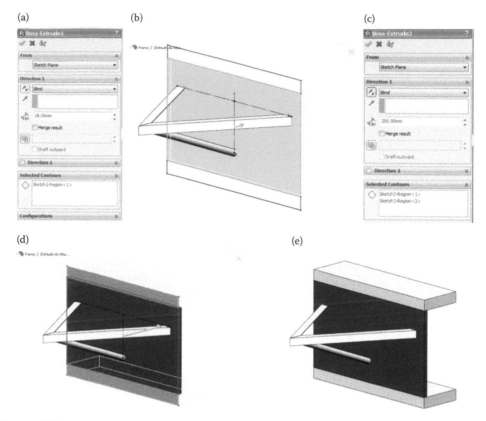

(d)   (e)

Figure 9.5

Development of CAD model of the structure – stage 9. (a) Boss/Extrude property manager – middle area extrusion. (b) Extrusion to 18 mm of the picked area. (c) Boss/Extrude property manager – side areas extrusion. (d) Extrusion to 200 mm of the side areas to form the C-channel outline. (e) The ready supporting plate.

coloured in pink in Figure 9.6b and c. The software automatically suggests which segments of the structural members to discard and which to keep.

Further, we have to model the horizontal surface plate. To do so, we have to build the plane of that plate.

11. Development of the horizontal **Plane4**.

Features → Reference Geometry → Plane (⬦) → OK

The plane is developed to coincide with the upper side of the angle iron profiles. It is fully defined by picking the upper face of one of the horizontal profiles and pushing the coincide button (Figure 9.6d). A view of **Plane4** is shown in Figure 9.6e.

12. Drawing of **Sketch4** in **Plane4** (Figure 9.6f). The sketch outlines the horizontal surface plate. It is a half segment of an ellipse, with half-axes that coincide with the constructive lines of the spatial frame. You can see the dimensions in the figure.

13. Defining the plate.

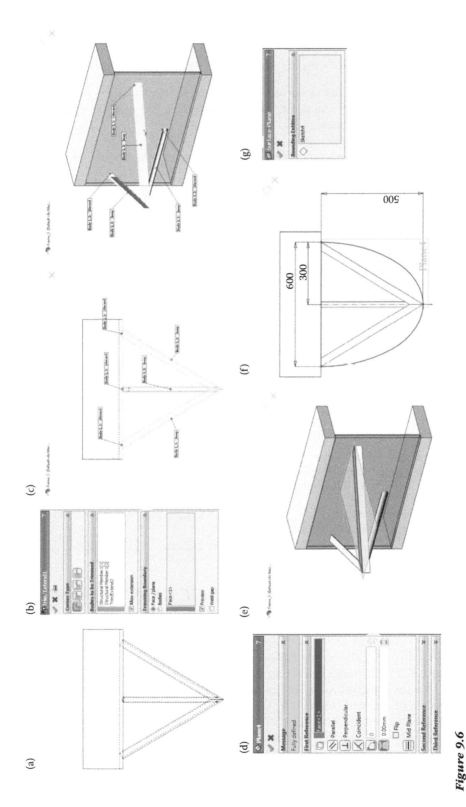

Figure 9.6

Development of CAD model of the structure – stages 10, 11, 12 and 13. (a) Interferences between the profiles and the supporting plate. (b) Trim/Extend property manager. (c) Trimetric view and top view of the discarded and kept body parts. (d) Plane4 property manager. (e) Plane4 view. (f) Sketch4. (g) Surface-Plane property manager.

At first, we activate the **Surface** tool (⬡) and then pick the **Planar Surface** (▣) command:

Surface → Planar Surface (▣) → OK

As **Sketch4** is still selected, its signature is automatically displayed in the window of the **Surface-Plane** property manager (Figure 9.6e). Thus, the area of **Sketch4** is transformed into a horizontal plate.

The CAD model of this complex structure is ready.

Regarding our intention to perform a static FEA of this model, we admit that it unites

- A body that will be treated as a solid body and which is the supporting C-channel form
- Three structural members made of two different types of ISO profiles, which will be modelled using beam FEs
- A horizontal plate, which is to be treated as a shell

Therefore, all types of structural components that the software adopts and analyses are included in this model.

This section helped us to exercise the modelling techniques using Weldments and Surface tools. We developed a complex structure, which combines 1D, 2D and 3D bodies. Despite its complexity, the structure is graded and saved as a part file (*.sldprt).

We exercised once more the modelling techniques using specific tools such as

- Weldments tool, which is used to model 1D parts (structural members as components of the complex spatial frame)
- Surface tool, which is used to model a 2D shell (the horizontal plate in the analysed structure)
- Features tool, which is used to model the complex 3D body, consisting of three extruded objects (C-channel supporting plate)

9.2 STATIC FINITE ELEMENT ANALYSIS OF THE STRUCTURE

A static finite element analysis (FEA) of the created structure will be made.

The developed CAD model (⬡) combines three different types of bodies (Figure 9.7a). The materials of all components are selected from the library of **SW Materials**.

We have modelled a 3D frame consisting of three beam members (⬡, Figure 9.7b). It is assumed that they are made of aluminium profiles. 1060 Alloy is an aluminium alloy with modulus of elasticity equal to 69 GPa, Poisson's ratio of 0.33, mass density of 2700 kg/m^3 and tensile strength of 68.94 MPa. The aluminium alloy is a ductile material, that is, its tensile and compressive strengths are assumed to be equal. The factor of safety of the 1D components is calculated regarding the yield strength of the alloy, which is 27.57 MPa. Rigid connections at both ends of each structural member are set through the **Apply/Edit Beam** property manager.

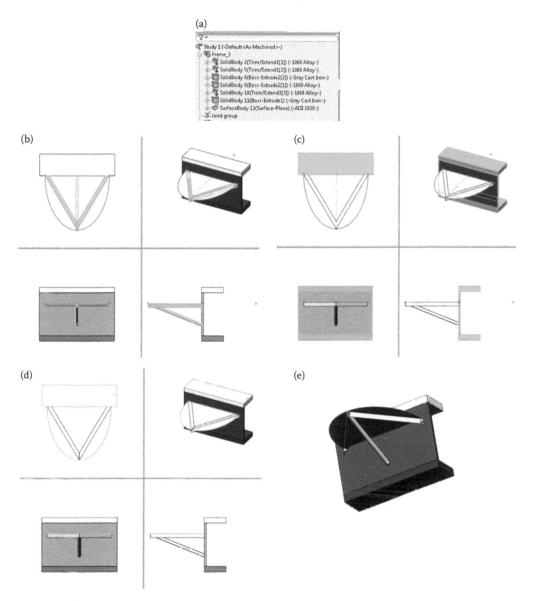

Figure 9.7
Different types of components in the analysed structure. (a) Different types of structural components. (b) Views of all beam members. (c) View of all solid bodies. (d) View of surface plane. (e) Calculated joints.

The second group of bodies consists of three solid bodies, which form the C-channel support (⌧, Figure 9.7c). They are made of Gray Cast Iron, which is a brittle material. Its modulus of elasticity is 66.178 GPa and its Poisson's ratio is 0.27. The mass density is equal to 7200 kg/m³. The compressive strength of the material (equal to 572 MPa) is a few times larger than its tensile strength (equal to 152 MPa). The factor of safety of these structural members is calculated according to Mohr failure criteria.

The material of the horizontal surface shell (⬡, Figure 9.7d) is assumed to be AISI 1020 steel. Its material properties are modulus of elasticity equal to 200 GPa, Poison's ratio of 0.29, mass density equal to 7900 kg/m³ and tensile strength of 420 MPa. The thickness of the plate is 4 mm. When the static analysis is performed, thick-plate formulation is used. The last two shell properties are introduced through the **Shell Definition** property manager.

All joints (⬡, Figure 9.7e) are automatically calculated by the program. There are three end joints, coloured in green, which mark the boundary between the beams and the solid body, and one joint, connecting all beams, which is coloured in pink.

While performing an FEA of a structure uniting different types of components, the correct input of contact settings is crucial for obtaining the accurate results. Unfortunately, the software cannot assess the accuracy of the boundary conditions; thus, the user's experience is significant at that stage.

Contact settings describe the interaction between part boundaries that are initially contacting or that come into contact during loading. The contact functionality is available in assembly and **multi-body part documents**, which is the studied case. Discussing the introduction of contact settings in a multi-body part analysis, we direct our progress to the static analysis of an assembled structure (*.sldasm).

The **Connections** icon (⬡) appears above the **Fixtures** icon (⬡) in the **Simulation study tree**. After defining the contact settings, any further change in contact conditions requires re-meshing of the model. If the study is run after the modification of contact settings, the software re-meshes the model automatically.

The right-mouse menu for the **Connections** icon provides different contact options, some of which will be discussed in the following.

At first, it is important to know that multiple contact conditions can be specified for an entity. The software enforces them as follows:

- **Global contact condition** is to be used for all <u>touching entities</u> for which no component or local contact condition has been specified. To apply **Global contact condition**, we select the top level assembly or multi-body part. By default, the software automatically applies automatic bonding between all touching entities of the model. Even more, the software automatically defines interaction between solids, shells and beams in a mixed mesh. Bonding beams to shells or solid faces is done automatically for touching components, and the algorithms imbedded in the software automatically bond them for the following cases:
  - A face or edge of a shell with another shell (Figure 9.8a–d)
  - A face or edge of a shell with a solid (Figure 9.8e and f)
  - A face of a shell with a structural member (Figure 9.8g and h)

Shells refer to both sheet metals and surfaces. All bonding contacts get transferred to the mid-surface of the shell automatically. The user needs to manually define bonding for touching non-planar faces that are meshed with shell elements. Bonding between touching structural members with a sheet metal face is also automatically created and transferred to mid-surface shells.

SW Simulation applies the following types of global contacts:

- **No Penetration** – This contact type is available for static analysis. It prevents interference between **Set 1** and **Set 2** entities but allows gaps to develop (Figure 9.9a). This is the most time-consuming option to solve.

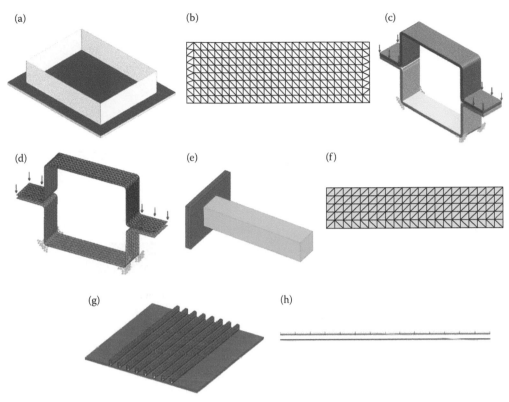

Figure 9.8
*Different types of bonded contact between different touching entities (SW Simulation help).
(a) Surface edge touching a sheet metal face. (b) Side view of mesh (zoomed). (c) Sheet metal faces
touching each other. (d) View of Mesh at Mid-Surfaces. (e) Solid face touching a sheet metal face.
(f) Side view of mesh (zoomed). (g) Structural members touching a sheet metal face. (h) Side view
of mesh (zoomed).*

- **Bonded** – This type of contact is available for all types of studies that require meshing. The program bonds the **Set 1** and **Set 2** entities (source and target), which may be touching or within a small distance from each other (Figure 9.9b). Bonded entities behave as if they were welded. The mesh does not have to be compatible. If the mesh is compatible, the program merges coincident nodes along the interface (Figure 9.9c); otherwise, it applies constraint equations internally to simulate bonding. For a study with a mixed mesh (used automatically by program), as is the discussed example, the user can bond **Set 1** entities (vertices, edges, faces, beam joints and beams) to **Set 2** (faces). Vertices, edges and faces can belong to shells or solids. The behaviour of the bond depends on whether the source entity belongs to a solid or a shell. If the source entity belongs to a shell or a beam, the bond acts like a rigid connection where the original angle between the shell or the beam and the solid is maintained during deformation. If the source entity belongs to a solid, the bond acts like a hinge, where the original angle between the shell and the solid is not necessarily maintained.

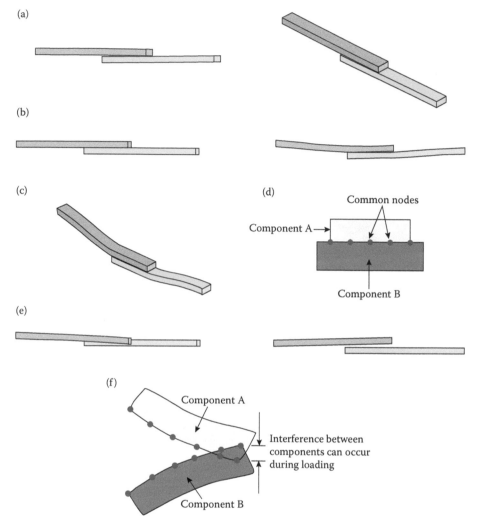

Figure 9.9
Different types of contact between parts (SW Simulation help). (a) No penetration contact. (b) Bonded contact. (c) Bonded contact. (d) Bonded contact with compatible mesh. (e) Allow penetration contact. (f) Allow penetration contact with incompatible mesh.

- **Allow Penetration** – This type of contact is available for static and other analysis. The program treats **Set 1** and **Set 2** faces as disjointed. For static studies, the loads are allowed to cause interference between parts. Using this option can save solution time if the user is convinced that the applied loads do not cause interference. The program meshes the common areas with an incompatible mesh.
- **Component contact** (🖳) conditions override **global contact** conditions. Modifying or adding a contact condition requires re-meshing the model. The component contacts are specified between the selected components (🦶) of the assembly. The components can be picked from the fly-out **Feature Manager** design tree or from the **Graphics area** directly. The available options for the **Component contact** depend on the study type. The component contact can be

- **No Penetration** – Selected components or bodies do not penetrate each other during simulation, regardless of their initial contact condition. **Surface to surface** contact formulation is applied for **No Penetration** contact.

  If **No Penetration component contact** option is selected, we can introduce friction between touching entities. To obtain accurate results, it is recommended that the friction coefficient ($\mathcal{L}$) be smaller than 0.5. If you do not know the exact value of the friction coefficient, you can set it to 0.2.

- **Bonded (No clearance)** – Selected components or bodies behave as if they were welded during simulation.

  If the **Bonded component contact** option is selected, we can choose either the **Compatible Mesh** or the **Incompatible Mesh** at the areas of contact. The program creates a compatible mesh on initially contacting areas. If the mesh is compatible, the program merges coincident nodes along the common interface. If the mesh is incompatible, the program meshes each component independently. If meshing fails with the compatible mesh option, the incompatible mesh option can help the meshing process to succeed. In general, the compatible mesh option produces more accurate results in the bonded regions.

- **Allow Penetration** – Selected components or bodies can penetrate each other during the simulation. The **Allow Penetration** option overrides other existing component contacts.

- **Local contact conditions** ($\blacksquare$) override global and component contact conditions. They can easily be defined using the **Contact Sets** property manager. The contact sets can be identified manually through the **Manually select contact sets** check button by further selecting pairs of faces and creating contact sets or automatically through the **Automatically find contact sets** check button. Using the automatic option urges us to be more careful because sometimes the automatic detection tool may not find all the contact sets that we want or it may find extra contact sets that we do not want.

The type of the local contact is visible only when the **Manually select contact sets option** is picked. We can choose among **No Penetration**, **Bonded**, **Allow Penetration**, **Shrink Fit** and **Virtual Wall**. As **No Penetration**, **Bonded** and **Allow Penetration** have already been discussed, we will focus on the remaining two options:

- **Shrink Fit** – This type of contact is available for static and nonlinear studies only. **Shrink Fit** refers to fitting an object into a slightly smaller cavity. Due to normal forces that develop at the interface, the inner object shrinks while the outer object expands. The amount of shrinkage or expansion is determined by the material properties as well as the geometry of the components. This is a local contact condition.

- **Virtual Wall** – This type of contact is available for static studies only. It defines the contact between the **Set 1** entities and a virtual wall defined by a target plane. The target plane may be rigid or flexible. If the virtual wall is chosen to be flexible, we can define the wall axial stiffness ($\boxplus$) and the wall shear stiffness ($\boxplus$). Further, the user can define friction between the **Set 1** entities and the target plane by assigning a non-zero value for the friction coefficient.

The **Local Contact** conditions can be applied to different entities, combined in two sets.

Set 1 (⬚) entities can be **Faces**, **Edges** and **Vertices**. For complex structures, a mixed mesh is applied, and if bonded type of contact is selected, we can pick **beam joints** (⬚) or **beams** (⬚). The last option is suitable for connecting beams to shell or to solid faces.

For **Set 2** (⬚), faces must be selected. If we define a **Virtual Wall** contact, then we select the **Target Plane** (⬚) as **Set 2**.

Entities of **Set 1** and **Set 2** must belong to different components, bodies, etc.

If the local contact sets are to be found automatically by pushing the **Automatically find contact sets** check button, we can select between two options: **Touching faces**, which sets contacts between selected touching faces, and **Non-touching faces**, which sets contacts between faces within the specified minimum (⬚) and maximum (⬚) distances for selected components. Further, we have to select the touching components and the software itself:

- Evaluates the possible contact sets among selected bodies or even the entire assembly
- Finds contacts of a single component or body with the neighbouring components
- Finds face pairs between the selected components that meet the specified criteria

As a result, the software displaces the possible contact sets, depending on the previously selected options.

Despite our choice of how to input the local contact sets, automatically or manually, if **No penetration** contact is preferred, we have to choose the type of contact at an FE level. It can be the following:

- **Node to Node** (Figure 9.10a): This contact type prevents interference between the source and target faces but allows them to move away from each other to form gaps. The program selects the candidates for source and target faces internally between **Set 1** and **Set 2** and creates coincident nodes on them. It requires compatible meshes for **Set 1** and **Set 2** entities and allows faces only for **Set 1**. This contact is available for initially touching contact faces only. **Node to Node** formulation supposes faster calculations than **Node to Surface** and **Surface to Surface**, but it is the least accurate for general structural problems with sliding or large rotations. The accuracy of the results depends on the loading, being best if the two faces are pressed against each other without much sliding or relative rotations. The accuracy reduces as the loading causes large sliding or rotations. For such problems, we use the **Node to Surface** or **Surface to Surface** options and activate large displacements.
- **Node to Surface** (Figure 9.10b and c): This contact type prevents interference between the source and target faces but allows them to move away from each other to form gaps. This contact type does not require that the faces be initially touching. **Node to Surface** formulation does not require a compatible mesh between source and target faces, that is, compatible meshes for **Set 1** and **Set 2** entities. For each node on the source, the software assigns one or more element faces on the target. We can pick vertices, edges and faces as **Set 1** entities. Although **Surface to Surface** contact is more accurate in general, the **Node to Surface** option gives better results if the contact area between the two faces becomes very small or reduces to a line or point.
- **Surface to Surface**: This contact prevents interference between the source and target faces during loading but allows them to move away from each other to form gaps. It is more general than **Node to Node** and **Node to**

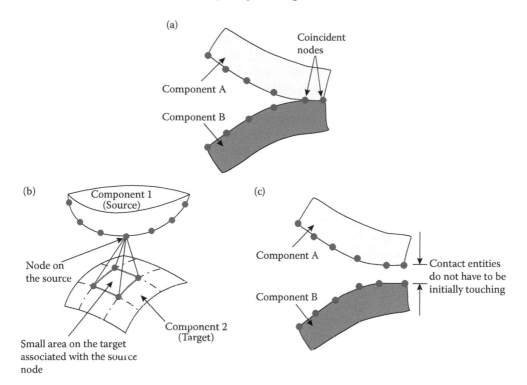

Figure 9.10
Different types of contact at finite element level (SW Simulation help). (a) A possible deformation of two faces that were initially touching by Node to Node contact. (b) A possible deformation of two faces with incompatible mesh. (c) A possible deformation of two faces that were not initially touching and use incompatible mesh.

Surface contacts. It is suitable for complex contacts with general loading. It does not require a compatible mesh between source and target faces, that is, between **Set 1** and **Set 2** entities. The program selects the candidates for source and target faces internally. The contact allows only faces as source and target entities. In most cases, **Surface to Surface** gives more accurate results but requires more time and resources. It is not recommended when the area of contact between the source face and the target face becomes too small or reduces to a line or a point during deformation. In such cases, we rather use the **Node to Surface** option.

After making a detailed review of the types, commands and options and the ways to input contact sets on the model, we can summarise the following general guidelines for specifying contact conditions:

- Check the interference between components before meshing. To detect interference in an assembly, you must click **Tools, Interference Detection**. The **Treat coincidence as interference** option allows you to detect touching areas, which are the only areas affected by the global and component contact (🖳) settings.
- For defining a local contact, use the **Contact Set** (🖳) to define the connection between solids, shells and beams.

- If no contact conditions are specified, the software assumes that all parts are bonded at their initially **touching entities**. All other entities are free.
- Specify global, component and local contact conditions efficiently to define the problem. For global and component contact, there is no need to select specific entities since they apply only to initially touching areas. The global contact is used to define the most commonly desired condition and then to override it by specifying component and local contact wherever needed.
- The **Find contact sets** property manager helps in finding and defining contact pairs between solids without having to select faces manually.
- After editing or defining contact conditions, the model must be re-meshed.

The most common contact condition is **bonding**. Bonding ensures the continuity of the model and transfers loads between two entities. The user can bond a face or an edge to any other face or edge. The meshes of the bonded entities do not have to be compatible. **Contact Set** (⌸) defines bonding conditions between solids, shells and beams. Bonding with a compatible mesh gives better results, but it can cause meshing to fail for some assemblies. Using the **incompatible mesh** can help in meshing such models. The entities do not have to be touching. The program bonds entities that are not too far apart or slightly interfering entities. Bonding is achieved by merging nodes when the mesh is compatible or by using multi-point restraints internally when the mesh is not compatible. Bonding incompatible meshes can generate local stress concentrations in the bonded areas. When bonding solid faces through the global contact condition, the program generates a compatible mesh on the touching areas and merges the nodes (Figure 9.11a). For example, meshing the model in Figure 9.11b with the default global contact setting (**Bonded**) bonds the whole circular face cylinder to the plate. If the cylinder is connected to the plate by welding its edge only, then set the **Global Contact** to **Free** and then bond the face of the plate to the edge of the cylinder by defining a local contact set. If a small clearance exists, the global contact settings are irrelevant, but local contact sets can bond the cylinder's face or edges to the plate (Figure 9.11c).

Regarding our new knowledge, we will discuss the contact sets in the structure.

We activate the menu by right clicking on the **Connections** icon (⍢, Figure 9.12a). The second and third lines in that menu include the **Component contact** (⌸) and **Contact set** (⌸) commands that we will use in defining contact conditions.

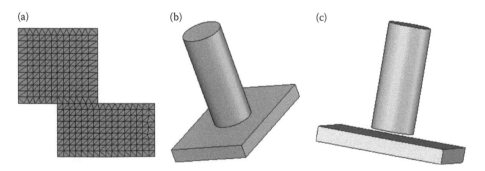

(a)   (b)   (c)

Figure 9.11
Different types of contact between two solid bodies (SW Simulation help). (a) Compatible mesh between two contacting solid bodies. (b) Bonded global contact between the cylinder and the face. (c) Use of a local contact set to bond the bodies while keeping the clearance between them.

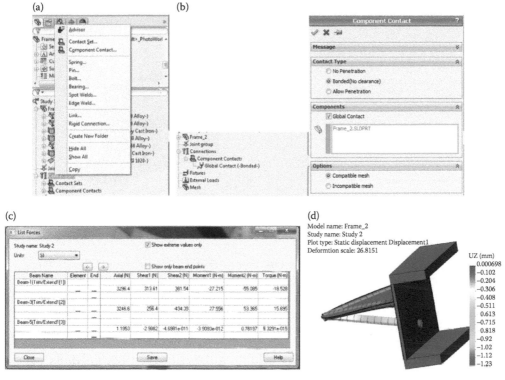

Figure 9.12
Calculation of the model at a global contact level. (a) Right click on Connections menu. (b) Global contact and Global Contact property manager. (c) Extreme values of beam forces. (d) Horizontal displacement plot (UZ).

As previously mentioned, the structure combines three types of components: solid bodies, shells and beams. Therefore, we have to consider all types of contacts allowed by the software.

The first type of contact to be discussed is the **Global** contact. This is the top-level contact, and it is applied to the entire model (**Frame_2**) by default. The software automatically chooses this contact to be **Bonded** (Figure 9.12b).

If we mesh and run the study at that stage, considering the input fixtures and loads, the program will provide us some results. Extreme values of the beam forces are given in Figure 9.12c, yet their accuracy is questionable. According to the list, there are almost no forces transferred to the pipe. Even more, if we plot the horizontal displacements of the frame components (UZ plot, Figure 9.12d), we will see that the pipe penetrates the plate, which cannot be true in the real structure.

As a result, we can conclude that it is obligatory to add component and local contacts.

The first to be introduced is a **component contact** (🖳, Figure 9.13a) at the joint common to all structural members (the pink joint in Figure 9.13b). After clicking on the **Component Contact** (🖳) on the **Connections** (🗍) right click menu (Figure 9.12a), the **Component Contact** property manager opens (Figure 9.13a). We select the type of the contact to be bonded, but instead of checking the **Global Contact**, we pick the three structural members by clicking on them in the **Graphics area** (Figure 9.13b). As a result, their signatures automatically appear in the blue **Components** window. We select to use

(a)   (b)

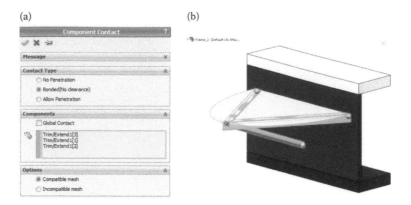

Figure 9.13

*Calculation of the model while only global contacts are set. (a) Component contact property man-
ager. (b) Graphic area view.*

the compatible mesh and click on the **OK** green mark to confirm our choice. Thus, a
connection at a component level between the three structural members is defined.

Yet, this is not enough to obtain correct results from the analysis. While the defined
component contact is optional, as the software has already established a bonded contact at the joint, it is obligatory to input a local contact between the vertical plate of the
C-channel and the pipe.

This time, we pick the **Contact Set** (▣) from the right click **Connections** menu
(Figure 9.12a), and our choice opens the **Contact Sets** property manager (Figure 9.14a).
The selected contact type is **Bonded**, and **Set 1** (set in the blue window) consists of the
three end (green) joints. In order to select them, we pick the **Joint** icon (▣). **Set 2** (set
in the pink window) includes the front face of the vertical solid plate (Figure 9.14b). The
software introduced this local contact on a separate line in the **Connections – Contact
Sets** group of the **SW Simulation analysis tree** (Figure 9.14e). The program uses
an icon, drawing the contact as a vertical solid body and a horizontal beam attached to it
(▣), to describe the contact in the **SW Simulation analysis tree**.

The last optional local contact, which duplicates the initial bonded settings of the
global contact defined by the software, is the contact between the angle iron profiles
and the bottom side of the horizontal shell. Of course, we can omit it, but when it
comes to the fact that the correct identification of different contacts in complex structures is crucial, it is recommended to consider it. The software will overwrite the
new boundary conditions over the existing global contacts. Sometimes it is better to
duplicate some of the contact sets instead of omitting them, which has been proved by
our initial calculations. We introduce this local contact (▣) through the **Contact Sets**
property manager (Figure 9.14a).

The selected contact type is **Bonded**. To enable selection of the structural members, we click the **Beam** (▣) icon. **Set 1** (set in the blue window) consists of the
two angle structural members (blue profiles in Figure 9.14d), whereas **Set 2** (set in the
pink window) includes the horizontal shell (Figure 9.14d). The program sets a new line
in the group of **Contact Sets** to describe the new contact. It uses an icon, drawing a
horizontal plate supported by beams (▣), to describe it.

By setting these additional local contact sets, the software provides correct results.

Before starting the analysis, it is better to clarify our choice of boundary conditions
once more. Regarding the input through the **Apply/Edit Beam** property manager

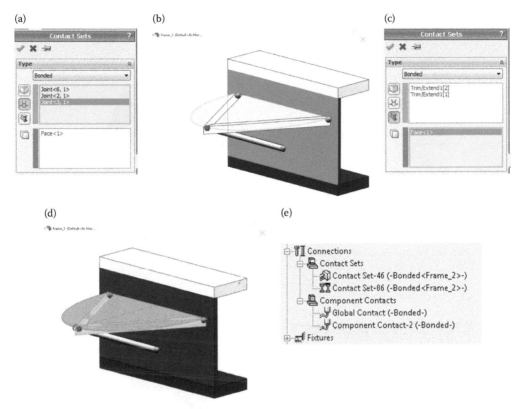

Figure 9.14
Calculation of the model at a local contact level. (a) Contact Sets property manager – joints' local contact. (b) Picked contacted face of the solid plate. (c) Contact Sets property manager – structural members' local contact. (d) Picked components of Set 1 and Set 2 of the local contact between the beams and the horizontal plate. (e) All input contacts as they are signified in SW Simulation analysis tree.

rigid boundary conditions at the ends of the structural members, the static scheme of each structural member is assumed rigidly fixed at both ends of the beam.

The next stage of the development of the static FE model is the identification of the fixtures. Different cables, located inside the channel support, do not focus our attention and are not discussed here. The C-form is fixed to a rigid vertical wall. Therefore, we assume **Fixed Geometry** fixtures at the back faces of the C-form (Figure 9.15) and introduce them through the path

Fixtures (, rick click of the mouse) → Fixed Geometry ( )

The structure will be studied under the following static loads:

- **Gravity** – it is inevitable for every structure. For the successful input of gravity, the density of each material must be defined. Figure 9.16a shows the options and values in the **Gravity** property manager, and the path through, which to start is

External Loads (right click) → Gravity ( ) → OK ( )

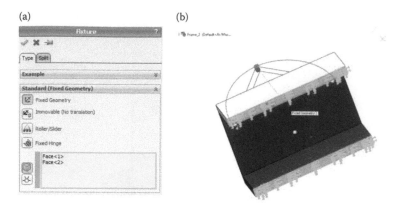

Figure 9.15
Applying fixtures to the model. (a) Fixture property manager. (b) Back faces of the solid body, where Fixed Geometry boundary conditions are applied.

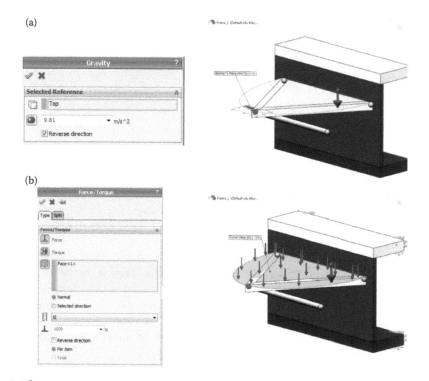

Figure 9.16
Applying the loads at the model – scenario 1. (a) Input of Gravity load. (b) Input of Force load, as a uniform load on the top side of the plate – scenario 1.

- **Force** – let us suppose that this force is the impact of a load of an object whose mass is approximately 100 kg disposed at the upper side of the horizontal plate. The total value of the force is about 1000 N. We assume that it is uniformly spread across the entire area of the plate – scenario 1 (Figure 9.16b).

External Loads (right click) → Force (⬇) → OK (✓)

If we want to be more precise in the input of the force load, we must consider the geometry of the bottom face of the object put above. It is supposed to be a cylinder with a diameter of 300 mm. We will study this option under scenario 2.

To start scenario 2, we duplicate **Study 1** by

Study 1 (the tab at bottom of the SW area, right click) → Duplicate → Study 2

Then we re-define the options of **Force** load by splitting the loaded area. To do so, we click the **Split** tab on the **Force** property manager and click the **Create Sketch** button (Figure 9.17a). The program automatically recognises the plane of the plate as a drawing plane. We sketch a circle of a diameter of 300 mm and coordinates of the centre (0, 200 mm), according to Figure 9.17b. Then we split the face of the plate to a circle and a surrounding area (Figure 9.17c and d). Finally, we input the force properties, including the circle area, where the force is applied (Figure 9.17e). As a result, the force is spread only across the outlined segment of the plate.

The next stage is meshing the structure. The software automatically recognises the type of the appropriate FEs for each structural member. For this case study, the FE model is a combination of solid FEs, shell elements and beam elements. Everything we have learned up to now, including the impact of the order of the used FEs, of the aspect ratio, of the size of the FEs, of the mesh control, etc. is applicable.

We start the solution of the case study by meshing and running scenario 1. We activate **Study 1** by clicking the corresponding tab at the bottom of the working area.

We see that after picking the **Study 1** tab, the software marks some of the contact sets between the angle profiles and the plate and the input **Force** as incorrect (❌, Figure 9.18a). The reason is the splitting of the plate into two separated areas. The easiest way to overcome that problem is as follows:

- Either to suppress the **Split Line 1** (⬜) in the **SW Design tree** (Figure 9.18b) by right clicking on the feature and picking the **Suppress** icon (⬇️) on the pop-down menu.
- Or by rolling the **End Line** to the previous and turning off the **Split Line 1** (⬜). Its visualisation below the **End Line** informs the user that in spite of its inclusion in the **SW Design tree**, this item is not active (Figure 9.18c).

The view of the result of each of the two options is shown in Figure 9.18d.

After deactivation of the **Split** feature, the program automatically applies the new features to **Study 1**, and as a result, the FE model is correct. We will not apply mesh control options at this stage. Curvature-based mesh is chosen. The density of the mesh is at the middle of the scale – minimal size of FE is set to 10 mm and maximal size to 30 mm (Figure 9.19a). The largest values of the **Aspect ratio** are below 4 (Figure 9.19b) and are calculated for FEs at the ends of the vertical support. Thus, we can assume that the meshing of the structure suits to our expected level of accuracy and precision of the results.

Finally, the **Run** command is started.

Some of the obtained results are systematised later when comparison among all studied scenarios is done.

To run correctly **Study 2** (scenario 2), we unsuppress the **Split Line** feature.

The mesh used in **Study 2** keeps the properties of the mesh in **Study 1**, in relation to the type, density and mesh control. The only difference is due to the existence of

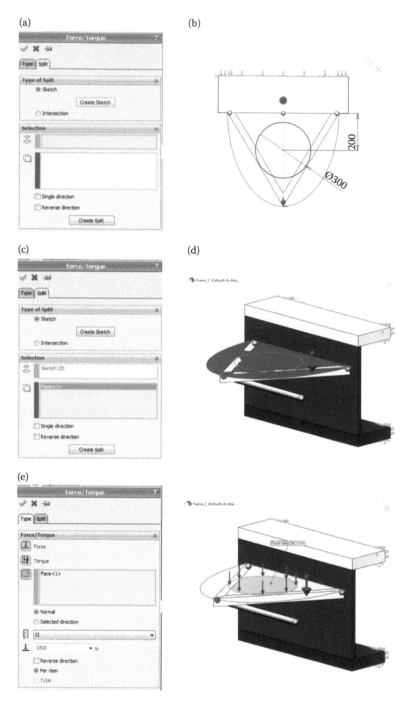

Figure 9.17
Applying the force to the model – scenario 2. (a) Force/Torque property manager – Split tab. (b) Sketch of the circle. (c) The input of the sketch line to split the violet face. (d) Graphic area view of the face to be split. (e) Input of the force at the split circle area.

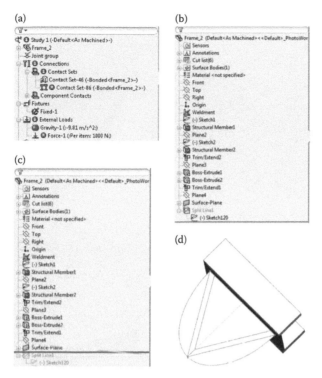

Figure 9.18
Deactivation of Split feature. (a) The incorrect SW Simulation analysis tree, including software massages. (b) The suppressed Split Line 1 in the SW Design tree. (c) Exclusion of Split Line 1 feature from the list of active commands. (d) Graphic area view after deactivation of the Split command.

a split area in the middle of the shell. The software automatically considers the circle outline (Figure 9.20).

To continue exploring the structure, we duplicate **Study 1** and **Study 2** to **Study 3** and **Study 4**, respectively. The last two scenarios differ from the previous ones in the static scheme of the pipe. We assume a hinge connection at **End 2** (⬤, blue end) and a rigid connection at End 1 (⬤, red end, Figure 9.21). The **Hinge** connection ensures

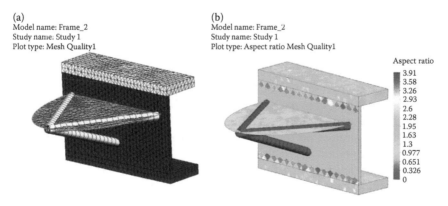

Figure 9.19
Meshing the structure – scenario 1. (a) Plot of the mesh. (b) Plot of the aspect ratio.

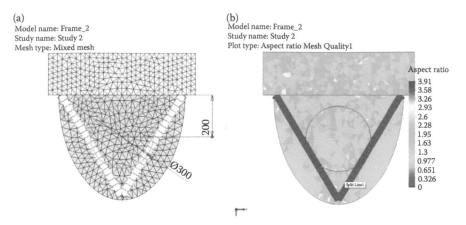

Figure 9.20
Meshing the structure – scenario 2. (a) Plot of the mesh – top view. (b) Plot of the aspect ratio – top view.

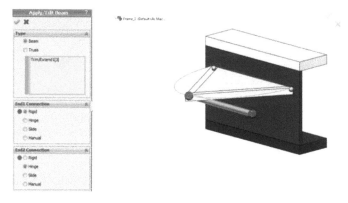

Figure 9.21
Definition of the new restraints of the pipe – scenarios 3 and 4.

that **End 2** can rotate freely and does not transfer any moments to the joint. The **Rigid** connection ensures continuity and fully defines the transfer of forces and moments. In the real structure, these connections can be interpreted as edge welding along the connection to the plate edge (End 1) and as a spot welding at a few points at the edge of End 2, for example.

Considering the previous remarks about the **Split Line** feature and its activation, we are ready to run these two scenarios and compare the final results.

Our last attempt focuses on entirely a new type of connection of the three structural members. It allows rotation of each member around the two axes located in its cross section. The new restraints prevent the angle profiles from rotating along their axes, while this torsional rotation is enabled for the pipe. The new studies are titled **Study 5** and **Study 6**. To develop them, we duplicate **Study 1** and **Study 2** and edit the member constraints according to the following instructions:

- The name of the component (right click to open the menu) → Edit definition → OK

This redefinition of the beam constraints is done for each beam, and the properties input through the **Apply/Edit Beam** property manager for every separate beam are given in Figure 9.22. While introducing the new restraints, we must be very careful about the colourful red-blue signature at the ends of the beams. After changing the restraints, and redefining the contacts and the external loads, we run the analyses.

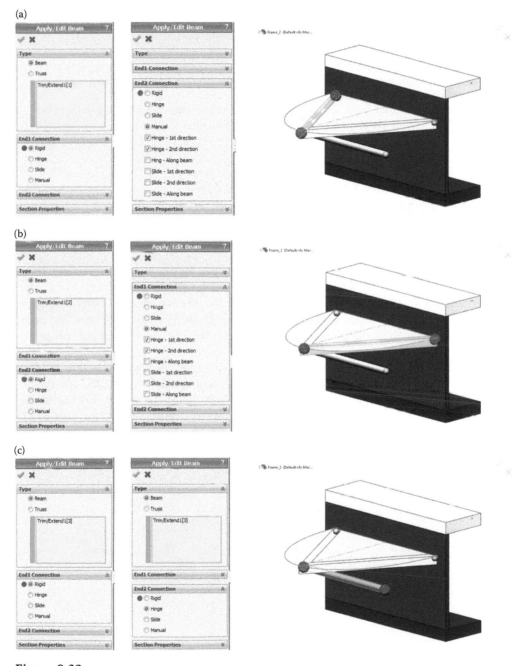

Figure 9.22
Definition of the new restraints of all 1D components – scenarios 5 and 6. (a) Restraints of the first angle iron profile. (b) Restraints of the second angle iron profile. (c) Restraints of the pipe.

In this section, we analysed a complex structure with 1D, 2D and 3D components. We focused our attention on setting different constraints and contacts. We discussed how to combine different types of FEs and the necessity of defining boundary conditions.

We exercised all previously discussed techniques in developing an FE model.

We learned how

- To define different contacts, including global contact, component contact and local contact sets
- To define a mixed mesh and to make all structural components work together and transfer loads and deformations among each other, etc.

9.3 COMPARISON OF THE RESULTS OF THE SIXTH DESIGN SCENARIOS

A brief comparison of all the obtained results is given in the following.

9.3.1 Definition of Stress Plots

When analysing an FE model with a mixed mesh, the **Stress Plot** property manager looks more complex. It combines the features of the property managers, which appear when a solid or shell structure is studied with the features of the 'beam' property manager.

In fact, the only more complicated and totally new sub-window is **Display** (Figure 9.23). The main point is that we have to select which type of stress to display. If we pick the **Solid and Shell** option, the software automatically limits the list of possible stress components (🔲) to normal stresses (SX, SY and SZ), shear stresses (TXY, TXZ and TYZ), principal stresses (P1, P2 and P3), von Mises stresses, etc., in which directions can be

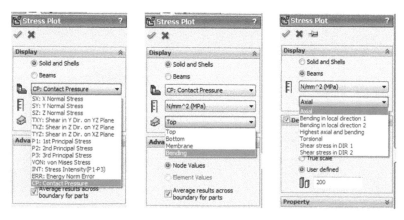

Figure 9.23
Stress plot property manager of a complex structure.

modified using appropriate reference geometry ( ). As far as the shell ( ) is concerned, we can choose the among the Top (total stresses = bending + membrane at the top face), the Bottom (total stresses = bending + membrane at the bottom face), the Bending (bending stress component) or the Membrane (membrane stress component). When the software plots the stress components for beams and we choose among different options, it displays either normal (axial; bending in local direction 1; bending in local direction 2; highest axial and bending) or shear stresses (torsional; shear stress in Dir 1; shear stress in Dir 2).

Some of the solid and shell plots are shown in Figures 9.24 through 9.26).

The extreme values for all case studies of the plotted stresses are systematised in Table 9.1.

We see that as far as the stress–strain state of the solid body and the shell is concerned, the shell is the more vulnerable component, especially while the circle bottom of the shell body put above is projected (Study 2 results in Figures 9.24 through 9.26). The results for all studies of that group are similar, including Study 2, Study 4 and Study 6. However, the precise adjustment of the loads ensures the accuracy of the FE model and outlines clearly the vulnerable zones of the shell – at the middle and at its connection to the vertical plate.

Further, the extreme values of the beam stresses are systematised in Table 9.2. They are listed through

Results (right click) → List Beam Forces → List Beam forces property manager (check Forces) → Select in the table SI Units and check Show extreme values only → Close

The largest are the stresses due to **Bending Dir 2** for the angle profiles and due to **Bending Dir 1** for the pipe. We must remember that the yield strength of the used aluminium alloy is 27.57 MPa, which guarantees FoS close to 1 and above.

All these stress values correspond to the beam plots shown in Figure 9.27.

The most significant is the difference between the stress distribution of **Study 1** and **Study 6**. The extreme stress values are influenced by two factors:

- **Introduction of the pressure load** – uniform distribution across the entire shell versus the circle in the middle
- **The connection and the end definition on the beam** – rigid beam ends connected to the pink joint in **Study 1** versus hinge beam ends in **Study 6**.

The second factor influences more strongly the stress distribution along the beams than their extreme values.

The software lists the stress values at both ends of each beam:

Results (right click) → List Beam Forces → List Beam forces property manager (check Stresses) → Select in the table SI Units and check Show only beam end points → Close

The software displays the results for **End 1** (the red end) of each beam in red and the results for **End 2** in blue (Figure 9.28a). However, sometimes it is difficult to merge all beams' ends in a common joint based on that signature, and the user must carefully consider the signature of the beam, particularly the start–end direction. Therefore, it is better to use the pink-green signature of the joints. The middle pink joint, for example, unites one red and two blue ends (Figure 9.28b).

The stress values, according to the pink-green joint signature, are given Table 9.3.

(a)

(b)

(c)

(d)

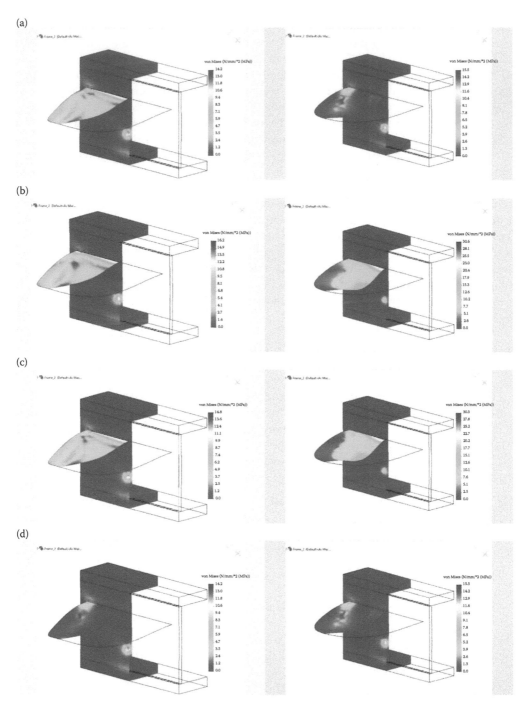

Figure 9.24

Von Mises (VON) stresses inside the solid body and the shell (MPa). All figures on the left are for Study 1 and all figures on the right for Study 2. (a) Plots of von Mises stresses inside the solid body and on the top of the shell. (b) Plots of von Mises stresses inside the solid body and at the bottom of the shell. (c) Plots of von Mises stresses inside the solid body and bending stresses at the shell. (d) Plots of von Mises stresses inside the solid body and membrane stresses at the shell.

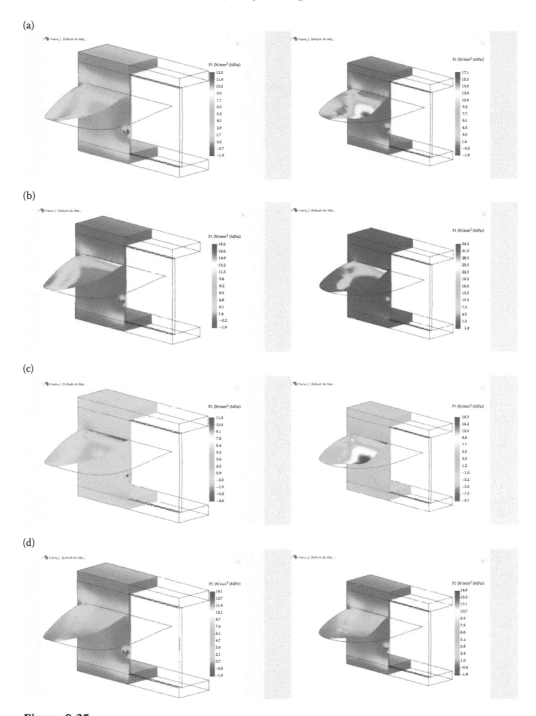

Figure 9.25
First principal (P1) stresses inside the solid body and the shell (MPa). All figures on the left are for Study 1 and all figures on the right for Study 2. (a) Plots of first principal stresses inside the solid body and on the top of the shell. (b) Plots of first principal stresses inside the solid body and at the bottom of the shell. (c) Plots of first principal stresses inside the solid body and bending stresses at the shell. (d) Plots of first principal stresses inside the solid body and membrane stresses at the shell.

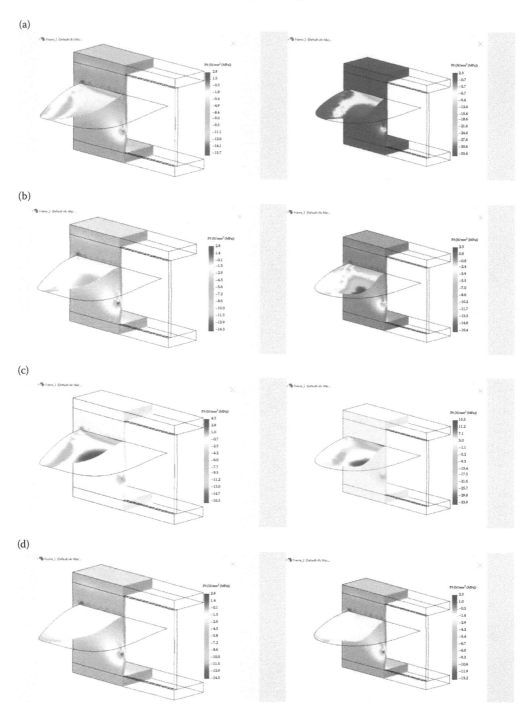

Figure 9.26
Third principal (P3) stresses inside the solid body and the shell (MPa). All figures on the left are for Study 1 and all figures on the right for Study 2. (a) Plots of third principal stresses inside the solid body and on the top of the shell. (b) Plots of third principal stresses inside the solid body and at the bottom of the shell. (c) Plots of third principal stresses inside the solid body and bending stresses at the shell. (d) Plots of third principal stresses inside the solid body and membrane stresses at the shell.

Table 9.1
Extreme Stress Values of Solid and Shell Structural Components

		Study 1	Study 3	Study 5	Study 2	Study 4	Study 6
von Mises Stresses (MPa)							
Top	Node			19,535			
	Stress	14.17	14.52	14.51	29.95	30.25	30.26
Bottom	Node		19,520			19,535	
	Stress	16.21	16.52	16.56	30.63	30.95	30.96
Bending	Node			19,535			
	Stress	14.84	15.19	15.19	30.29	30.60	30.61
Membrane	Node		19520			19490	
	Stress	14.16	14.47	14.49	15.53	16.02	16.01
First Principal Stresses (MPa)							
Top	Node		19,520			19,490	
	Stress	14.08	12.91	12.93	17.09	17.62	17.62
Bottom	Node			19,535			
	Stress	18.25	18.67	18.67	34.34	34.70	34.71
Bending	Node		16,820			19,828	
	Stress	11.87	12.14	12.14	16.32	16.45	16.45
Membrane	Node		19,520			19,490	
	Stress	14.07	14.37	14.40	14.87	15.34	15.33
Third Principal Stresses (MPa)							
Top	Node		19,532			19,535	
	Stress	−15.67	−15.99	−15.88	−33.56	−33.89	−33.90
Bottom	Node		16,204			19,828	
	Stress	−14.30	−14.47	−14.46	−16.39	−16.54	−16.55
Bending	Node			19,535			
	Stress	−16.46	−16.86	−16.86	−33.95	−34.30	−34.30
Membrane	Node		16,204			16,174	
	Stress	−14.30	−14.47	−14.46	−13.21	−12.61	−12.62

We see the zeros in all beams at the pink joint for **Study 5** and **Study 6**. They are due to bending in both directions and the introduced hinges. The same is the explanation of the zero stresses for **Beam 3** (the pipe) in **Studies 3** and **4**.

9.3.2   Definition of Plots of Inner Beam Forces

The next optional stage is plotting the diagrams of the inner beam forces. To start the procedure, the user can follow the path

Results (right click) → Define Beam Diagrams…

We have already discussed how the **Beam Diagrams** property manager (Figure 9.29) helps in drawing the beam diagrams. Beam diagrams are generated according to the local directions of each beam. We must choose the **Component** (🖳) among Axial

Table 9.2
Extreme Stress Values of Beam Structural Components

	Study 1	Study 3	Study 5	Study 2	Study 4	Study 6
Axial						
Beam 1	−2.831	−2.907	−2.909	3.633	3.848	3.847
Beam 2	2.712	2.807	2.810	3.661	3.860	3.858
Beam 3	−3.611	−3.601	−3.598	−3.126	−3.106	−3.106
Bending Dir 1						
Beam 1	5.469	5.552	5.580	3.068	3.182	3.173
Beam 2	6.119	6.208	6.230	4.091	4.232	4.229
Beam 3	5.579	4.524	4.518	5.186	3.945	3.946
Bending Dir 2						
Beam 1	16.391	16.755	16.782	15.508	16.062	16.057
Beam 2	17.182	17.510	17.552	16.469	17.012	17.004
Beam 3	2.780	3.012	3.01	5.168	2.430	2.430
Worst Case						
Beam 1	24.691	25.214	25.271	21.646	22.429	22.415
Beam 2	25.927	26.406	26.477	23.405	24.186	24.173
Beam 3	9.425	9.036	9.027	8.771	7.739	7.740

Force, Shear Force in Dir 1, Shear Force in Dir 2, Moment in Dir 1, Moment in Dir 2 or Torque and the **Units** (⊞). We can generate beam diagrams for all beams (by checking **All**) or for selected beams (by checking **Select** and picking the beams (🖱) in the **Graphics area**).

The beam diagrams are shown in Figures 9.30 and 9.31. The main difference between the diagrams of the two studies are at the common (the pink) joint. The extreme values of the inner forces for all diagrams are given in Table 9.4.

While the extreme values of the inner forces (stresses) are very important for the successful sizing of the beams, the values at both beam ends are significant when the connectors are designed. Even more, these values help the designers in calculating tapered beams. The inner force data for all beam ends are systematised in Table 9.5. The zeros correspond to the input hinges in Studies 3, 4, 5 and 6.

9.3.3   Definition of Displacement Plots

We will view the displacement plots on the deformed shape (Figures 9.32 and 9.33) through the path

Results (right click) → Define Displacement Plot...

The extreme values of the nodal displacements, including nodal rotations, are given in Table 9.6. They help the designer to assess the deformations in the structure and to compare them to the limits in the regulations.

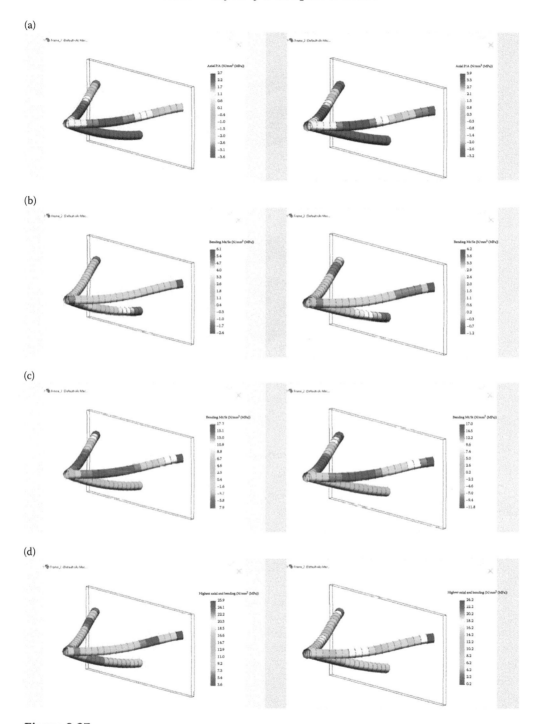

Figure 9.27
Plots of stresses inside the beams (MPa). All figures on the left are for Study 1 and all figures on the right for Study 2. (a) Plots of axial stress (P/A). (b) Plots of normal stress as a result of bending in local direction 1 (Ms/Ss). (c) Plots of normal stress as a result of bending in local direction 2 (Mt/ St). (d) Plots of total normal stress as a result of the highest axial and bending stress.

(a)

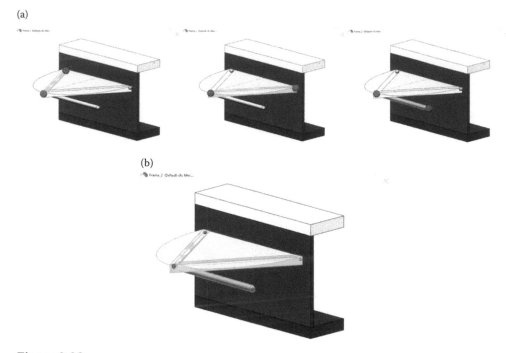

(b)

Figure 9.28
Different ways to display beam ends. (a) Red-blue signature at the end of each beam element. (b) Pink-green signatures of all joints in the structure.

While comparing the results, keep in mind that the vertical axis in this model is Y and the horizontal axes are X and Z.

9.3.4 Definition of Deformation Plots

Additional information about the stress–strain state of the structure can be provided by

Results (right click) → Define Design Inside Plot …

This plot helps in finding the most vulnerable zones. The slider constantly increases the percentage of the applied loading, and thus, the software shows those areas that carry the loads most efficiently and that constantly enlarge during the process. The provided information is intended to be used to reduce the material and to optimise the model's shape. The blue areas are loaded the most, whereas the translucent areas outline the model and the areas of the material, which can be discarded in a future optimisation. The pictures below show the structure state at 30% to 90% of the loading (Figure 9.34).

Regarding the plots, we can conclude that, if we need to make the structure steadier and suitable for exposing to higher loads, we better increase the thickness of the horizontal plate and re-design the connections between the shell and the vertical plate and between the pipe and the vertical plate than strengthen the

Table 9.3
Extreme Stress Values at the Beam Joints (Figure 9.28)

	Study 1	Study 3	Study 5	Study 2	Study 4	Study 6
Axial – Pink Joint						
Beam 1	0.724	−0.0857	0	−0.0273	0.0524	0
Beam 2	0.796	0.959	0.840	1.117	1.450	1.649
Beam 3	−3.606	−3.596	−3.593	−3.121	−3.101	−3.101
Axial – Green Joint						
Beam 1	−2.8314	−2.907	−2.909	−3.070	−3.184	−3.185
Beam 2	−2.627	−2.6882	−2.695	−2.845	−2.942	−2.941
Beam 3	−3.611	−3.601	−3.598	−3.126	−3.106	−3.106
Bending Dir 1 – Pink Joint						
Beam 1	2.329	−0.0454	0	0.0881	0.242	0
Beam 2	5.262	3.526	0	0.584	0.664	0
Beam 3	−2.442	0	0	−2.283	0	0
Bending Dir 1 – Green Joint						
Beam 1	5.469	5.552	5.580	3.068	3.182	3.173
Beam 2	6.119	6.208	6.230	4.092	4.232	4.229
Beam 3	5.579	4.524	4.510	5.186	3.945	3.946
Bending Dir 2 – Pink Joint						
Beam 1	3.757	0.0775	0	0.151	0.413	0
Beam 2	4.792	3.086	0	1.075	−1.213	0
Beam 3	2.780	0	0	5.168	0	0
Bending Dir 2 – Green Joint						
Beam 1	16.391	16.755	16.782	15.508	16.062	16.057
Beam 2	17.181	17.51	17.552	16.469	17.012	17.004
Beam 3	1.640	3.012	3.01	0.227	2.430	2.430
Worst Case – Pink Joint						
Beam 1	6.810	0.209	0	0.266	0.708	0
Beam 2	10.851	7.571	0.840	2.776	3.327	1.649
Beam 3	7.305	3.596	3.593	8.771	3.101	3.101
Worst Case – Green Joint						
Beam 1	24.691	25.214	25.271	21.646	22.429	22.415
Beam 2	25.927	26.406	26.477	23.405	24.186	24.173
Beam 3	9.425	9.036	9.027	8.318	7.739	7.740

beams. Another solution is to attach additional ribs or other strengthening devices to the plate.

The factor of safety plot shows another vulnerable zone – the connection of the angle profiles to the vertical plate (Figure 9.35). The FoS is minimal at that area and its values are

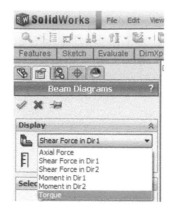 

Figure 9.29
Beam Diagrams property manager.

Study 1—minFoS = 1.064   Study 2—minFoS = 1.178
Study 3—minFoS = 1.044   Study 4—minFoS = 1.140
Study 5—minFoS = 1.041   Study 6—minFoS = 1.141

We viewed and compared the results for a few studied examples.

We paid attention to the necessity of defining contacts between different structural members. This is a step towards learning how to perform static analysis of more complex structures and assemblies.

We proved that every structure can be optimised with regards to certain constructive requirements. We studied the impact of the applied connections on the stress and displacements across the structure and proved the modification of connectors to be one of the easiest ways to reduce the stresses in the most vulnerable areas.

We explained how to use the Design Inside plot as a guide to constructive optimisation resulting in material reduction and better shape for the input restraints.

We learned

- How different contact sets and connectors impact the results, and consequently how to improve the stress distribution by modification of the connectors
- How to use the Design Inside tool as a guide in performing a structural optimisation resulting in material reduction

Analysis of that complex structure is the step crossing the gap between the analyses of parts and structural analyses of complex structures designed as assembled models.

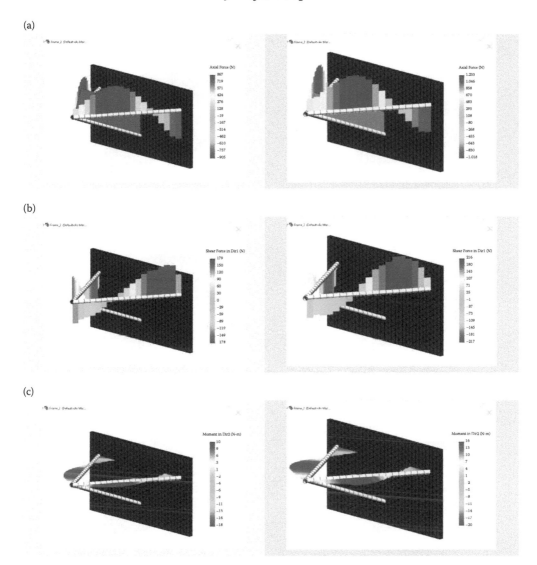

Figure 9.30
Beam diagrams of the studied frame – part 1. All figures on the left are for Study 1 and all figures on the right for Study 2. (a) Diagrams of axial force (N). (b) Diagrams of shear force in direction 1 (N). (c) Diagrams of moment in direction 2 (N m).

(a)

(b)

(c)

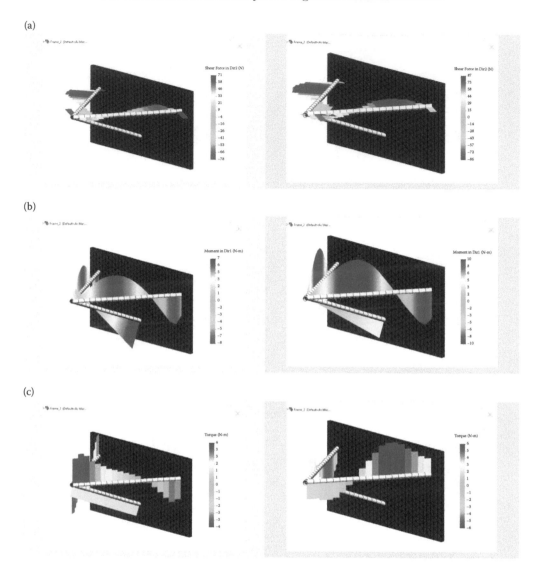

Figure 9.31
Beam diagrams of the studied frame – part 2. All figures on the left are for Study 1 and all figures on the right for Study 2. (a) Diagrams of shear force in direction 2 (N). (b) Diagrams of moment in direction 2 (N m). (c) Diagrams of torque (N m).

Table 9.4

Extreme Values of Inner Forces for All Beams

	Study 1	Study 3	Study 5	Study 2	Study 4	Study 6
Axial Force (N)						
Beam 1	905.02	929.19	929.88	−1160.6	−1229.8	−1229.7
Beam 2	−866.68	−897.08	−898.30	−1170.8	−1233.8	−1233.1
Beam 3	860.24	857.86	857.27	744.50	739.64	740.11
Shear Force in Direction 1 (N)						
Beam 1	178.50	181.94	182.15	209.19	216.78	216.8
Beam 2	179.49	182.39	182.67	209.10	215.62	215.54
Beam 3	−4.325	−8.974	−8.969	−13.745	−7.525	−7.527
Bending Moment in Direction 2 (N m)						
Beam 1	17.707	18.130	18.143	18.403	19.060	19.061
Beam 2	−18.256	−18.628	−18.667	−18.897	−19.510	−19.503
Beam 3	−3.520	3.815	3.812	−6.538	3.076	3.077
Shear Force in Direction 2 (N)						
Beam 1	77.044	78.382	78.901	109.32	−86.963	−86.969
Beam 2	78.003	71.517	71.625	83.639	86.078	86.049
Beam 3	19.917	11.234	11.22	18.426	9.792	9.800
Bending Moment in Direction 1 (N m)						
Beam 1	−7.798	−8.011	−8.009	9.258	9.588	9.592
Beam 2	−7.33	−7.506	−7.520	−9.157	−9.550	−9.545
Beam 3	−7.065	−5.729	−5.722	−6.547	−4.994	−4.998
Torque (N m)						
Beam 1	3.845	4.020	3.846	6.267	6.469	6.496
Beam 2	−3.949	−4.168	−4.590	−6.105	−6.286	−6.287
Beam 3	−1.503	0	0	−1.403	0	0

Table 9.5
Inner Forces at Beam Ends

	Study 1	Study 3	Study 5	Study 2	Study 4	Study 6
Axial – Pink Joint (N)						
Beam 1	231.52	−27.4	0	339.15	437.2	0
Beam 2	−254.44	−306.4	−268.6	−362.14	−461.82	−527.1
Beam 3	−859.04	−856.67	−856.07	−743.3	−738.44	−738.91
Axial – Green Joint (N)						
Beam 1	905.02	929.19	929.88	981.18	1017.7	1017.9
Beam 2	−839.77	−859.25	−861.4	−909.5	−940.31	−939.96
Beam 3	860.24	857.86	857.27	744.5	739.44	740.11
Shear 1 – Pink Joint (N)						
Beam 1	145.76	11.935	10.068	126.23	112.43	1.118
Beam 2	126.65	126.17	88.684	119.67	115.51	128.12
Beam 3	1.337	−8.9738	−8.969	10.757	−7.525	−7.527
Shear 1 – Green Joint (N)						
Beam 1	86.5	87.111	87.518	27.846	29.005	28.861
Beam 2	120.51	121.53	121.76	68.762	71.029	70.984
Beam 3	−4.325	5.986	5.981	−13.745	4.537	4.539
Bending Dir 2 – Pink Joint (N m)						
Beam 1	−3.214	0.0684	0	−1.589	1.300	0
Beam 2	2.300	1.373	0	0.965	−1.097	0
Beam 3	−3.520	0	0	0.290	0	0
Bending Dir 2 – Green Joint (N m)						
Beam 1	17.707	18.13	18.143	18.403	19.06	19.061
Beam 2	−18.256	−18.628	−18.667	−18.897	−19.51	−19.503
Beam 3	2.076	3.8145	3.812	−6.538	3.076	3.077
Shear 2 – Pink Joint (N)						
Beam 1	−2.714	3.5929	−53.247	−109.32	−73.938	12.295
Beam 2	78.003	35.533	−15.565	−29.122	−67.368	−51.18
Beam 3	−19.917	−11.234	−11.22	−18.416	−9.792	−9.800
Shear 2 – Green Joint (N)						
Beam 1	77.044	78.382	78.901	40.759	42.34	42.192
Beam 2	69.917	71.13	71.499	50.499	52.319	52.285
Beam 3	19.917	11.234	11.22	18.416	9.792	9.800
Bending 1 – Pink Joint (N m)						
Beam 1	0.1752	0	0	−2.815	−0.0010	0
Beam 2	−3.300	−2.3094	0	−0.272	0.0081	0
Beam 3	−3.092	0	0	−2.845	0	0

(continued)

Table 9.5 (Continued)
Inner Forces at Beam Ends

	Study 1	Study 3	Study 5	Study 2	Study 4	Study 6
Bending 1 – Green Joint (N m)						
Beam 1	5.532	5.7065	5.689	8.059	8.343	8.351
Beam 2	−5.279	−5.4192	−5.422	−7.445	−7.680	−7.678
Beam 3	−7.065	−5.729	−5.722	−6.547	−4.994	−4.998
Torque – Pink Joint (N m)						
Beam 1	−3.74	−0.3814	−0.544	3.245	2.933	2.751
Beam 2	−3.74	−3.768	−4.590	3.639	2.904	3.186
Beam 3	1.503	0	0	1.403	0	0
Torque – Green Joint (N m)						
Beam 1	−2.9623	−2.944	−3.003	2.665	2.731	0.126
Beam 2	−2.641	−2.612	−2.644	2.922	2.996	2.997
Beam 3	−1.503	0	0	−1.403	0	0

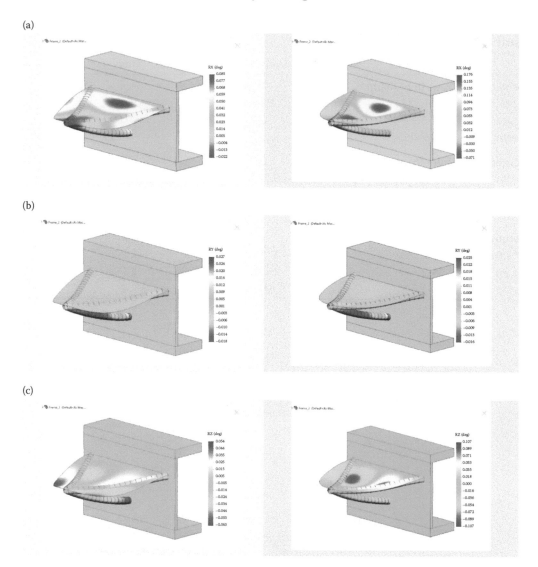

Figure 9.32
Plots of nodal rotations (degrees). All figures on the left are for Study 1 and all figures on the right for Study 2. (a) Plots of the nodal rotations around axis X (RX). (b) Plots of the nodal rotations around axis Y (RY). (c) Plots of the nodal rotations around axis Z (RZ).

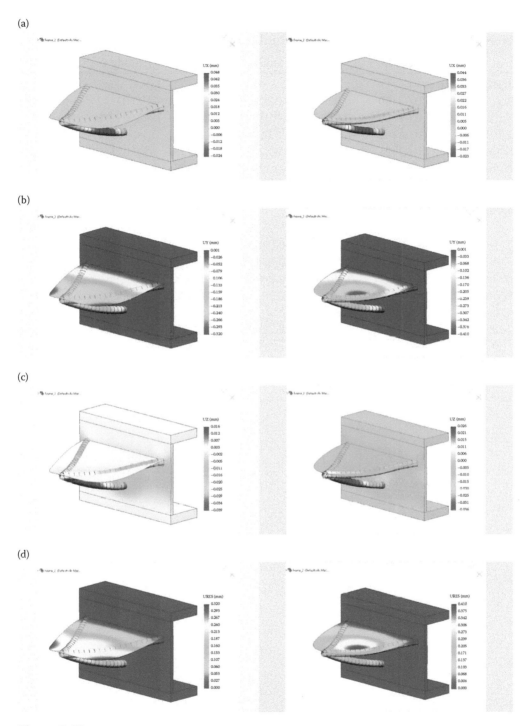

Figure 9.33
Plots of nodal displacements (mm). All figures on the left are for Study 1 and all figures on the right for Study 2. (a) Plots of the nodal displacements in direction X (UX). (b) Plots of the nodal displacements in direction Y (UY). (c) Plots of the nodal displacements in direction Z (UZ). (d) Plots of the total nodal displacements (UREZ).

Table 9.6
Extreme Values of Nodal Beam Displacements

	Study 1	Study 3	Study 5	Study 2	Study 4	Study 6
Algebraic Min						
RX (°)	−0.02159	−0.02675	−0.02627	−0.07071	−0.07142	−0.07142
RY (°)	−0.01765	−0.03750	−0.03743	−0.01641	−0.03270	−0.03271
RZ (°)	−0.06305	−0.06500	−0.06496	−0.10730	−0.10811	−0.10813
UX (mm)	−0.02421	−0.02483	−0.02482	−0.02551	−0.02166	−0.02166
UY (mm)	−0.31980	−0.32430	−0.32590	−0.41011	−0.41962	−0.41975
UZ (mm)	−0.03853	−0.03920	−0.03918	−0.03586	−0.03516	−0.03517
Algebraic Max						
RX (°)	0.08578	0.08766	0.08756	0.17596	0.17894	0.17900
RY (°)	0.02731	0.02840	0.02839	0.02528	0.02477	0.02477
RZ (°)	0.05403	0.05371	0.05458	0.10679	0.10761	0.10769
UX (mm)	0.04756	0.06794	0.06800	0.04400	0.05963	0.05925
UY (mm)	0.00112	0.00122	0.00122	0.00082	0.00082	0.00082
UZ (mm)	0.00164	0.00579	0.00579	0.02642	0.00595	0.00598
UREZ (mm)	0.31981	0.32428	0.32595	0.41006	0.41963	0.41976

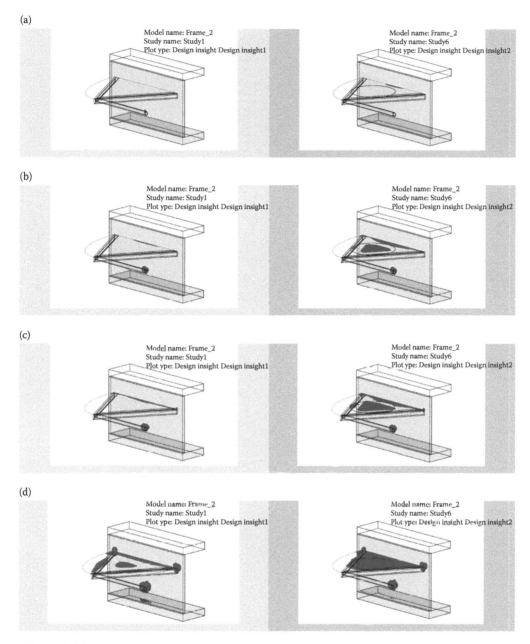

Figure 9.34
Design inside plots at a different level of loading. All figures on the left are for Study 1 and all figures on the right for Study 2. (a) 30% loading. (b) 50% loading. (c) 70% loading. (d) 90% loading.

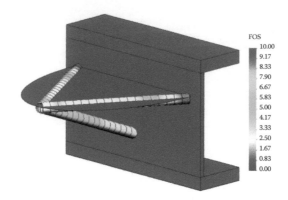

Figure 9.35
FoS plot for Study 1.

INDEX

Page numbers followed by f and t indicate figures and tables, respectively.

Index

Milton Keynes UK
Ingram Content Group UK Ltd.
UKHW020845141024
449569UK00003B/81